SECRETS OF THE LOST RACES

SECRETS OF THE LOST RACES

New Discoveries of Advanced Technology in Ancient Civilisation

Rene Noorbergen

Researched by Joey R. Jochmans

NEW ENGLISH LIBRARY/TIMES MIRROR

Dedicated to all those whose startling discoveries –
sparked the search for the existence of the lost races of men

---

First published in the USA in 1977 by The Bobbs-Merrill Company, Inc.

First published in Great Britain in 1978 by New English Library Ltd

© 1977 by Rene Noorbergen

First NEL Paperback Edition February 1980

NEL Books are published by
New English Library from
Barnard's Inn, Holborn,
London EC1N 2JR.
Made and printed in Great Britain by
William Collins Sons & Co. Ltd

45004563 **3**

CONTENTS

INTRODUCTION

With an impressive stack of science-fiction books under his arm, a college student walked reluctantly into my office only a few minutes after I had delivered a lecture on psychic phenomena.

His face was filled with utter perplexity as he pointed to his collection of books.

'Do *you* believe that we may have been the target of visitors from outer space?'

Leaning back in his chair, he waited impatiently for a reply. It was clear he did not expect a religious or philosophical answer, or any other reaction that would contradict what he had already accepted as truth.

He wanted confirmation.

I couldn't give it.

He was visibly puzzled. His eager gullibility had allowed him to soak up all the available information on the subject, yet he didn't feel free to advocate his new-found conviction unless he could support it with proof. But where was it? And, what's more, could it really be found?

My mind flashed back to other occasions when I had been placed in much the same position, and always my response had been the same.

His questions were undoubtedly representative of the vacillating opinions in the minds of millions, for books telling of intergalactic visitors to this planet are found in every major language of the world.

Enthusiasm for this by-product of the space age is limitless. Yet in our minds the fundamental questions remain unanswered.

- Did 'they' really visit us in ancient times?
- Is our technology a mere shadow of what 'they' once taught us?
- Are 'they' perhaps still in communication with our civilization, or have 'they' abandoned us – forever?
- Is it possible that the unidentified ancient artifacts attesting to a super-technological society belong to our own historical development?
- Can it be that we have regressed instead of evolved? Have we

7

lost more than we have gained?
- Are we now again approaching the level of advancement and sophistication that led to a historical downfall of the human race?

This book is an attempt to supply a reasonable answer for these questions, based on the numerous discoveries that have lain un-evaluated in stately museums and musty libraries.

The conclusions are startling and even frightening, for much that transpired in the dawn of history is lost beneath the rubble of our past.

Have we indeed been visited by 'them', or should we look to the achievements of our own race for the answers?

I trust that the recorded facts will lead to a solution of the riddle of time.

RENE NOORBERGEN
February 1977
Collegedale, Tennessee

CHAPTER 1

The End of the Beginning

Sophisticated as we may be in our endeavours to create new things and expand scientific knowledge, we have made the twentieth century both an age of development and an age of confusion.

And where it will end, no one knows.

We have lost something, and that something may well be a crucial part of our heritage.

Our slot in the eons of unrecorded time has in fact become an age of slanted information. With ever-increasing frequency, new hypotheses are tested, current scientific theories modified, and new formulae constructed, but all with the aim of proving previously accepted theories; and the bits of disconnected information that continually pierce the walls of scientific complacency are often ignored.

Our quest for knowledge *and* our memories have become one-sided, and this is precisely our problem.

In our modern society we have a name for everything, and each newly acquired fact is shoved into its own little niche, in the hope that one day it will fit somewhere into our still-developing society. We seem to have arrived at the point when we feel that we humans are the final recipients of all previously discovered knowledge. This, we tell ourselves, is the 'Age of Aquarius', the 'Age of Enlightenment'. Without saying it in so many words, we tend to believe that within the present generation is embodied all knowledge accumulated thus far, and that without us the future would look awfully dim.

But are we *really* all that smart?

For years we have been living with the theory of evolution, the theory of relativity and other scientific concepts, but while relativity is a theory that rests on valid scientific principles, our supposed development from protoplasm to ape to man is still the subject of much unsupported discussion, and consequently man stubbornly seeks for new ways to explain what thus far remains unexplainable. We need a fresh look at the world around us, for even though science has brought us from the kerosene

9

lamp to atomic power, we are steadily confronted with discoveries that somehow do not fit into the accepted theories of our development.

For more than a century, orthodox historians essentially have dealt with only a single body of historical facts – facts that meet the requirements of their preconceived hypothetical framework, telling them that man of today is the result of an evolutionary process that has brought us upwards both intellectually and physically from a lower order of beings. Millions of years are involved in this hypothetical view of history, and even though the historians cannot draw the curtain of recorded history back further than 6000 years, they steadfastly stick to their theories, for their programmed minds simply will not accept any other explanation for man's technological and cultural development.

Theirs is a one-sided view of a many-sided problem, for in order to achieve a total concept of history, two other bodies of facts must be taken into consideration – ooparts and the historical Bible.

What are ooparts? you wonder.

For the past thirty years, there has been a steadily increasing number of historical and archaeological discoveries made at various sites around the world, which, because of their mysterious and highly controversial nature, have been classified as 'out-of-place' artifacts – thus the name ooparts. The reason for this designation is that they are found in geological strata where they shouldn't be, and their sudden appearance in these layers of ancient dirt has baffled the minds of many a trained scientific observer. They emerge from among the remains of the treasured past sans evidence of any preceding period of cultural or technological growth. In many cases, the technical sophistication of the ooparts extends far beyond the inventive capabilities of the ancient peoples among whose remains they were discovered.

There is little doubt that these artifacts are also out of place in *theory*, for in no way do they conform to what is accepted as a part of the development of the human race. There is simply no time slot to accommodate them. Is it possible that the muted voice of the unknown past is trying to get through to us? Is it possible that these ooparts bear testimony to the existence of a super-civilization of human origin and development at some distant point in the history of mankind?

Sounds far out, doesn't it? Yet what are the chances of its being so? Is it conceivable that there may have been a period in human history when there was (a) civilization(s) that was (were)

10

comparable or possibly more advanced than our own twentieth-century society?

Today's bookstores are plentifully stocked with books written by persons trying to connect the ooparts with visits of beings from other galaxies who supposedly toured this planet more than 10,000 years ago, leaving behind proof of their transgalactic visitation. The major difficulty in accepting this theory is that none of the ooparts is composed of material unknown on earth, and their technological make-up conforms with the development of our own modern civilization. A closer look at the strange artifacts now suggests that the ooparts originated in a man-made civilization – one that antedated known history – one that attained an elevated degree of development, but was destroyed to such an extent by a devastating catastrophe in the distant past that only a few remnants of its science and technology survived among the inferior cultures that succeeded it in history.

But where do we find indications that a catastrophe of this magnitude occurred?

There is one possible explanation for the emergence of these 'remnants', but that would require an acceptance of the Genesis story of the Deluge – a flood that is reported to have obliterated mankind from the earth in a disaster so destructive, so complete, that one shudders when attempting to recollect it.

It is this third aspect of the historical account of man that may contain the facts needed to solve the oopart mystery. If we turn to the Bible as the third source of our historical knowledge, we find that the book of Genesis gives a full account of the annihilation of an entire civilization by means of the most destructive flood the world has ever experienced. This civilization flourished during what has been termed the antediluvian era – the period between the earliest recorded Biblical history and the worldwide catastrophe.

According to the Genesis account, the antediluvian people were highly knowledgeable, being the first to develop agriculture, animal husbandry, construction, architecture, political organization, metal-working, the abstract arts, mathematics, chronology and astronomy. What's more, while the Genesis record tells us that there were altogether only ten generations of antediluvians, they developed the majority, if not all, of the basic elements of civilization by the sixth generation. Now if the antediluvians, beginning with the raw earth, were able to master the arts of civilization in the first six generations of their existence, we may well wonder to what degree they further developed and refined

11

those arts in the remaining four prior to the Deluge.

Orthodox historians of course often disregard the significance of the historical books of the Old Testament and with them the story of the Flood and the antediluvian civilization. Yet the Genesis account of the Deluge may be the only answer to the enigma of the ooparts.

According to this story, the major force in the destruction of the advanced civilization of the antediluvians was the Flood, and in Chapter 7 of Genesis it is described as a violent cataclysm:

'The same day were all the fountains of the great deep broken up . . . And the waters prevailed exceedingly upon the earth, and all the high hills . . were covered.' The result was the death of all human beings, except those aboard the ark: 'And all flesh died that moved upon the earth . . . every man.' The convulsions of nature not only buried the antediluvian people but also utterly destroyed their technological achievements, for this tremendous upheaval was certainly enough to obliterate any form of machinery or construction.

'That's what the *Bible* says,' is a comment often overheard, 'but we all know something like a universal flood never really happened.' In the years I devoted to Deluge research, attempting to uncover extra-Biblical confirmation of a universal Deluge, I have encountered this judgement many times, and more often than not it is the result of total unawareness of all the evidence.

It was in 1947 that I first became acquainted with a group that was setting out for Mount Ararat in eastern Turkey to search for the legendary ark of Noah – the ship that is said to have carried the only surviving family of antediluvians across the waters of the Flood. If it were not for other sources beside the Biblical account of the catastrophe, the endeavours might not have sparked the imagination of agnostics; but it was the copiousness of secular information supporting the concept of an all-destroying Flood that intrigued them.

Universal Deluge Traditions

Over the years approximately 80,000 books pertaining to the Deluge have been written, in no less than seventy-two languages – and these are only the ones that are traceable through the card indexes of the great libraries of the world. Many more have undoubtedly been written but never catalogued. The majority of these are concerned with the archaeological and geological aspects of the Flood, and not with the legends and folklore underlying the basic history of the ancient civilizations; yet these tales are nevertheless of the utmost importance, as their very existence among widely separated tribes is what generally might be expected if the Flood was indeed a universal one.

The famed German researcher Dr Johannes Riem stated in the introduction to his impressive work on Flood legends, *Die Sintflut in Sage und Wissenschaft:*

'Among all traditions there is none so general, so widespread on earth, and so apt to show what may develop from the same material according to the varying spiritual character of a people as the Flood traditions. Lengthy and thorough discussions with Dr Kunnike have convinced me of the evident correctness of his position that the fact of the Deluge is granted, because at the basis of all myths, particularly nature myths, there is a real fact, but during a subsequent period the material was given its present mythical character and form.'[1]

His position is one that is echoed by many other researchers. The noted nineteenth-century Scottish geologist Hugh Miller, an ardent collector of the world's most haunting traditions, writes:

'There is, however, one special tradition which seems to be more deeply impressed and more widely spread than any of the others. The destruction of well-nigh the whole human race, in an early age of the world's history, by a great deluge, appears to have impressed the minds of the few survivors, and seems to have been handed down to their children, in consequence, with such terror-struck impressiveness that their remote descendants of the present day have not even yet forgotten it. It appears in almost every mythology, and lives in the most distant countries and among the most barbarous tribes . . .'[2] Yet among even these forgotten races of the human family [Indians of the Orinoco] he found the tradition of the Deluge still fresh and distinct; not confined to a single tribe, but general among the scattered nations of that great region, and intertwined with curious addition, suggestive of the inventions of classic mythology of the Old World.

Dr Riem's book had one other important feature: a world map showing the various areas where Flood traditions have been located. Most of the traditions were found in Asia and on the North American continent, but Australia, Europe, Africa and the South Sea Islands also had their own individual traditions.

Are the common denominators in these traditions significant enough that we can regard them as different versions of the same event?

13

Asia, particularly China, harbours some of the most striking Deluge traditions. For example, it is told that a tremendous flood of devastating force occurred there in approximately 2300 BC. According to this story the flood, caused by an overflow of the great rivers, was stopped by the eventual swelling of the sea. The Chinese hero escaped the destruction, together with his wife, three sons and three daughters. But that isn't all: additional legends found on the mainland of China contend that all Chinese are direct descendants of an ancient ancestor called Nu-wah, who was known for having overcome a great flood.

A fascinating aspect in connection with this Chinese flood story is that ancient Chinese writing contains words that can be traced *only* to 'Nu-wah' and the Flood. The Chinese word for 'righteousness', for example, is a combination of the pictorial symbol for 'lamb' placed over the ideogram for 'myself'. Theologians tell me that this is apparently related to Noah's desire to justify himself in the eyes of his God, as shown by the burnt offering he made after disembarking from the ark.

Dr E. W. Thwing, a researcher who has spent many years in China investigating the Noah account, comments:

'The Chinese have records and traditions of a great flood. And it is a curious fact that the word used for "ship" as printed in Chinese books and papers today is the very ancient character, made up of the picture of "boat" and "eight mouths", showing that the first ship was a boat carrying eight persons.

'In looking over some old books of ancient stories and traditions,' he continues, 'I found a story about the ancestor "Nu-wah". Interesting enough, "Nu" means "woman" and "wah" is "flowery." '[3] It seemed that the ancient one was considered to be a female ancestor. Further examination of the symbols, however, revealed two small "mouth" pictures placed *beside* the name, indicating that *not* the meaning but the *sound* of the characters was important, pointing towards a male ancestor named Nu-wah, an ancient man who escaped the wrath of the gods in a boat.

Early Chinese history further supports the idea of a Noachian involvement in the birth of China. It is mentioned that the destruction of the world by a flood was caused by Jung-ku, but the reconstruction was accomplished by Nu-wah. Three legendary men then succeeded Nu-wah, and even though at various times they are referred to as heroes, sovereigns or monarchs, they bridge the gap between the Ancient One and the first three Chinese

14

dynasties, the Hsia, the Shang and the Chou. Can it be that their Nu-wah and the Noah of the Bible are the same?

The name Noah has survived for thousands of years, through a multiplicity of stories, even though ofttimes it evolved into a slightly different spelling, depending on the letter symbols used. Such is the case with the Hawaiian legend of Nu-u, the Righteous Man. The island people believe that long after the creation of the

(*Left*) The Chinese word for "ship" is a combination of the symbols for "mouth" and "eight", which suggests that the ark was a boat that carried eight persons. (*Right*) The Chinese regard Nu-wah as their ancestor and hero. The characters which make up his name are used not for their meaning but for their sound. This strengthens the belief that the Chinese Nu-wah (Noah?) could indeed have been the ancestor of the Chinese.

world, the first man had become so wicked that his god, Kane, decided to destroy him and the earth on which he lived; but, being weary of creating, he decided to allow the one righteous man Nu-u and his family to escape his anger by building a great canoe with a house on it. Kane then instructed Nu-u to take his wife Lili-Nu-u, their children and all the animals he wanted on the boat to await the great flood.

When the rains came and the waters rose and the oceans merged, the Waa-Halau drifted for days on end, and while it drifted all of mankind was destroyed. Finally, after leaving the rainbow as a token of his eternal forgiveness, Kane made the waters subside and told Nu-u and his three sons to repopulate the earth.

More than thirty Flood legends have been discovered in the Orient, and the Indonesians can lay claim to possessing some of the most interesting.

'The Battaks of Sumatra say that when the earth grew old and dirty, the Creator – whom they call Debata – sent a flood to destroy every living thing. Debata was angry. The last human pair had taken refuge, not in an ark, but on the top of the highest mountain, and the waters of the deluge had already reached their knees when Debata, the Lord of all, repented of his resolution to make an end of all mankind.

'Magnificently picturesque legends have grown up among the natives of Engano, an island to the west of Sumatra, and among the Sea Dyaks of Sarawak in Borneo. The Bugi-speaking Toradjas of the Central Celebes tell of a flood which covered the highest mountain, leaving bare only the tip of Mount Wawom Pebato. This time no lucky pair escaped. Instead, the only living creatures to survive the flood were a pregnant woman and a pregnant mouse.'[4]

North America too has its share of Deluge legends, forty-six of which deal with the subject of a universal flood. Sherman Coolidge, an Arapaho, tells the following tale, highly revered by his tribe:

'Long ago, before there was any animal life on earth, the entire surface of the planet was covered with water, except the top of one high mountain. Upon this mountain sat a lone Arapaho, poor, weeping, and in great distress. The Great Spirit saw him and felt sorry for him, and in his pity sent three ducks down to the poor Indian. The Arapaho ordered the ducks to dive down into the waters and bring up some dirt. The first and second tried, but after remaining under water for a long time, each returned without any dirt. Then the third went down and was gone so long that the surface of the water where he disappeared had become still and quiet. The Arapaho believed his duck to be dead, when suddenly she returned to the surface with some dirt in her bill. As soon as the Arapaho received this bit of dirt, the waters began to subside.

'In a short time the waters had receded so far that they could not be seen from the top of the highest mountains, but this Arapaho, who was endowed with supernatural wisdom and power, knew that they surrounded the earth, even as they do to this day. The Arapaho, who had been saved by the ducks, then became the sole possessor of the land. He made the rivers and made the trees to grow and then the buffaloes, elks, deer, and all the trees and bushes and all other things that can be grown by planting seed in the ground.'[5]

In the western part of the United States, the Athapascan Indians have a tradition which tells about the gods repairing the skies as they are about to fall. In the aftermath, torrential rains engulfed the earth below.

'Every day it rained; every night it rained. All the people slept. The sky fell; the land was not. For a very great distance there was no land. The waters of the oceans came together. Animals of all kinds drowned. Where the waters went, there were no trees. There was no land. Water came, they say. The waters completely joined everywhere. Trees and grass were not. There were no fish or land animals or birds. Human beings and animals alike had been washed away. The wind did not blow through the portals of the world, nor was there snow, nor frost, nor rain. It did not thunder, nor did it lightning. Since there were no trees to be struck, it did not thunder. There were neither clouds nor fog, nor was there sun. It was very dark.

'Then it was that this earth with its great, long horns got up and walked away down this way from the north. As it walked along through the deep places, the water rose to its shoulders. When it came up into shallower places, it looked up. There is a ridge in the north upon which the waves break. When it came to the middle of the world in the east under the rising of the sun, it looked up again. Then where it looked up will be a large land near to the coast. Far away to the south it continued, looking up. It walked under the ground. Having come from the north, it travelled far south and lay down. Nagaitsche, standing on earth's head, had been carried to the south. Where earth lay down, Nagaitsche placed its head as it should be and spread grey clay between its eyes on each horn. Upon the clay he placed a layer of reeds and then another layer of clay. In this he placed upright blue grass, brush, and trees. "I have finished,' he said. "Let there be mountain peaks here on its head. Let the waves of the sea break against them."'[6]

And the earth was made new.

An almost limitless number and variety of universal Flood legends can be found in nearly every corner of the globe, but most significant is the fact that they all describe basically the same event.

'All these traditions have been modified through the ages,'

comments Alfred Rehwinkel in *The Flood*, one of the standard works dealing with the Deluge. 'They have been influenced by the customs of the various people and by the environment in which they are found and thus have taken on local colour and sometimes extravagant and fantastic proportions, so that the kernel of truth in many cases is seriously obscured. And yet, when stripped of the accretions which have accumulated as they were handed down from father to son through the generations, the essential facts of the great catastrophe are easily discernible. *There is almost complete agreement among them all on the three main features:* 1. There is a universal destruction of the human race and all other living things by water. 2. An ark, or boat, is provided as the means of escape. 3. A seed of mankind is preserved to perpetuate the human race. To these might be added a fourth, which, though not occurring in all the traditions, occurs very frequently, namely that the wickedness of man is given as the cause of the Flood.'[7]

Patriarchs, Gods and Kings

Other information that is equally as important as the Flood legends and traditions is concealed in the genealogies of the antediluvian patriarchs as listed in the book of Genesis. Critics have ridiculed this list of pre-Flood giants, for the Bible ascribed fantastic lifespans to these men. Methuselah, the oldest, is supposed to have lived 969 years, and the lives of the other patriarchs also far exceeded those of our generation.

Critics are prepared to discredit this entire section of Genesis, for they say these lifespans do not belong to the realm of possibility, but for now we are more interested in the names and the number of patriarchs than in their longevity. It seems reasonable that if the Bible story is accurate, Noah and his sons would have carried with them a fundamental knowledge of both the history and the technology that existed in the year prior to the Flood, and this history would undoubtedly include the number of antediluvian 'kings' – the leaders who ruled from the beginning of their time until the moment of destruction. It is also obvious that these names, much like the history of the Deluge itself, will have experienced linguistic modifications as they passed into the different languages that developed in various parts of the world. Another point to be considered is that, depending on the value these developing nations placed on their religion, politics or social status, these early patriarchs may have been given the

18

rank of king or god. If this is an acceptable hypothesis, history must have other lists of ten patriarchs – perhaps under other designations.

We don't have to look far for such a list, for both the Egyptians and the Babylonians had lists of ten 'antediluvian kings':

BIBLICAL PATRIARCHS	EGYPTIAN GODS	CHALDEAN KINGS
Adam	Ptal	Alorus
Seth	Ra	Aloparus
Enos	Su	Almelon
Cainan	Seb	Ammenon
Mahalaleel	Osiris	Amegalarus
Jared	Set	Daonus
Enoch	Hor	Aedorachus
Methuselah	Tut	Amempsinus
Lamech	Ma	Otiartes
Noah	Hor	Xisuthros

Whereas Noah is the hero of the Biblical Flood story, Xisuthros is the celebrated survivor of the similar Chaldean account. Upon comparing the list of the prehistoric Chaldean kings with the pre-Flood patriarchs, we find a fascinating parallel that again supports our basic premise. The name of the third king, Almelon, means 'a man'; the name of the third patriarch is Enos, which means 'moral, weak mankind'. The fourth name on the Chaldean list, Ammenon, means 'craftsman'; Cainan, the name of the forth Sethite patriarch, also means 'a craftsman'. Aedorachus, the seventh Babylonian king listed, has the connotation 'bearer of divine revelations, he to whom the secrets of heaven and earth are revealed'; Enoch, his Biblical counterpart, has a name meaning almost the same thing. The eighth king, Amempsinus, is given in the original Sumerian listing as Sukarlam, a name which in part bears a resemblance to that of Lamech, the ninth patriarch. Lastly, the tenth name, Xisuthros, according to Berosus, the high priest of the Babylonian temple Bel-Marduk, was the name of the hero who saved mankind from the terrible destruction. In the Sumerian royal lists, Xisuthros is called Utnapishtim, and following his name is written, 'After this, the Flood overwhelmed the land.' The famed Gilgamesh epic tells basically the same story, but since it was written on clay tablets several thousand years before Moses wrote the Noachian account in the

19

book of Genesis, sceptics have often raised the question of whether Moses borrowed the facts from the Babylonians, thereby making his story a new version of the same tragedy.

After a diligent comparison between the Babylonian and the Genesis stories, Merril F. Unger, author of *Archaeology and the Old Testament*, points out that:

a) Both accounts state that the Deluge was divinely planned.
b) Both accounts agree that the impending catastrophe was divinely revealed to a hero of the Deluge.
c) Both connect the Deluge with the wickedness of the human race.
d) Both assert that the hero of the Deluge was divinely instructed to build a huge boat to preserve life.
e) Both tell of the deliverance of the hero and his family.
f) Both indicate the physical causes of the Flood.
g) Both specify the duration of the Flood, although they differ in the reported time.
h) Both name the landing place of the boat.
i) Both tell of the sending forth of birds at certain intervals to ascertain the decrease of the waters.
j) Both describe acts of worship by the hero after his deliverance.
k) Both allude to the bestowing of special blessings upon the hero after the disaster.

Rehwinkel writes in *The Flood:*

'According to both accounts, the Flood is brought on because the earth was full of violence . . . In both cases dimensions of the ship are given, though differing in details. In both cases representatives of all animals are taken into the ark. In the Babylonian account the Flood lasts seven days. In the Bible narrative the embarkation takes seven days. In both cases a raven and a dove are sent forth from the ark. The Babylonian accounts add a swallow . . . The rainbow of Genesis is represented by the great jewels of Ishtar. In both there is a covenant guaranteeing that no world Flood is ever to come upon the earth again to destroy it.'[8]

The accumulated Flood legends and traditions tell a tale of a terrifying disaster, with sudden torrential rains, devastating waterspouts, and the agonized cries of drowning men and animals piercing the air. The story speaks of horrendous storms, cataclysmic earthquakes, and that terror-driven dash by man and beast to reach the safety of the ark, only to be dragged under by the relentless force of the onrushing waters.

Outside of the great ship, man ceased to exist. Flying reptiles, towering dinosaurs and the enticing beauty of the antediluvian world perished from the face of the earth. The destruction was complete – and sudden. Today palaeontologists are still puzzling over the fact that myriads of early life remains are embedded in the sprawling rock formations of the postdiluvian earth, and there simply is no way to account for them other than with the Flood story.

The unprecedented disappearance of entire civilizations and multitudes of diverse plants and animals must have left traces which awaited their discovery by an investigative mind.

But where are these traces?

Combining the pertinent facts, we conclude that the Flood was undoubtedly accompanied by violent winds that first swept the turbulent waters to the lower levels of the land masses, then continued swirling upwards until the peaks of the highest mountains were completely covered. It was a catastrophe that made the earth tremble and shake; never before or since has a calamity of such proportions and magnitude touched the globe on which we live. The upheaval was so terrific that today the greatest mountains on earth – the Rockies, the Andes, the Himalayas, the Alps – still bear the telltale signs of seashells and other evidence of ocean life that existed thousands of years ago. Their silent voice predates the records of the Egyptians and Babylonians by many years, but precisely how many years is a subject of scientific debate. Evolutionists believe there was a gap of millions of years; Deluge geologists claim it is only a matter of thousands. In any case, a global flood of this proportion unquestionably deposited vast amounts of sediment on the bottom of the newly formed water masses, and since this flood has been considered universal, it should be possible to discover remnants of such deposits at a variety of locations. *And this is exactly what's happened!* It has been scientifically estimated that over 75 per cent of the earth's surface is sedimentary in nature, with some areas having more and others less sediment. Testing has shown that while the United States has prodigious sedimentary deposits, centring in California and the Colorado plateau, India has the deepest sedimentary deposit found thus far: 60,000 feet deep!

The evolutionist theory that slow erosion was followed by gradual accumulation over a period of millions of years hardly

21

seems reasonable. Geologist Dr H. G. Coffin, of the Geoscience Research Institute in Berrien Springs, Michigan, writes in his book *Creation:*

'Such processes as gradual submergence and the slow accumulation of sediment by erosion seem inadequate to account for the great quantities of water- and wind-deposited materials. Adjacent areas do not provide sufficient material for decomposition on such a scale. But a flood of sufficient extent to cover all land, and a storm of great violence that stirred roiled water or soft mud is sufficient to account for the transport of vast amounts of sedimentary material over great distances, and the filling in of depressions irrespective of the height or extent of adjacent landscapes.'[9]

Faced with such criticism, scientists are willing to admit – though reluctantly – that perhaps there are exceptions to their 'gradual deposit theory'. The Rockies, for example, certainly do not fit the scientists' often preposterous claims. There, well-preserved water ripple marks and a countless number of trilobites, together with other delicate fossils without a sign of disintegration, are among the startling features that strongly suggest that they were laid down not in a slow and gentle manner but rather abruptly and suddenly, as if by a great and terrible catastrophe.

Forget the mistaken idea that silence has no voice. It does. The ancient bones speak with clear voices, and prehistoric animals tell their tales from the glass enclosures of the museums; even the voices of tiny fossil shells and small vertebrates found imbedded in the rocks are forceful.

Researcher Immanuel Velikovsky, who is highly controversial because of his unorthodox views, has examined the presence of fish in sedimentary rock, and his conclusions fully support the catastrophe theory:

'When a fish dies, its body floats on the surface or sinks to the bottom and is devoured rather quickly, actually in a matter of hours, by other fish. However, the fossil fish found in sedimentary rock is very often preserved with all its bones intact. Entire shoals of fish over large areas, numbering billions of specimens, are found in a state of agony, but with no mark of a scavenger's attack.'[10]

Many discoveries bear this out. It is estimated that eight hundred thousand million fish skeletons have been uncovered in the Karroo formation in South Africa.

Geologist Hugh Miller, writing about the Devonian rocks which cover most of the British Isles, comments that 'at this period in our history, some terrible catastrophe involved in sudden destruction the fish of an area at least a hundred miles from boundary to boundary, perhaps much more. The same platform in Orkney as at Cromarty is strewed thick with remains, which exhibit unequivocally the marks of violent death. The figures are contorted, contracted, curved; the tail in many instances bent around the head; the spines stick out; the fins are spread to the full as in fish that die in convulsions.'[11] The area described by Miller is not small; it covers at least ten thousand square miles of territory bearing all the evidence of having been exposed to a major destructive force. Describing a spot closer to home, Harry S. Ladd of the United States Geological Survey says that 'more than a billion fish, averaging six to eight inches in length, died on four square miles of bay bottom.'[12]

In speaking of the disappearance of the dinosaurs, Edwin H. Colbert once remarked, 'The great extinction that wiped out all of the dinosaurs, large and small, in all parts of the world, and at the same time brought to an end various lines of reptilian evolution, was one of the outstanding events in the history of life and in the history of the earth . . . It was an event that has defied all attempts at a satisfactory explanation.'[13]

'It is as if the curtain were rung down suddenly on a stage where all the leading roles were taken by reptiles,' says George Gaylord Simpson, one of the most respected men in palaeontology, 'especially dinosaurs, in great number and in bewildering variety, and rose again immediately to reveal the same setting but an entirely new cast, in which the dinosaurs do not appear at all, other reptiles are mere super-numeraries and the leading parts are all played by mammals.'[14]

Having researched the Flood story for many years, I have files on the subject that are bulging with information about this unexplainable event. For instance, a fascinating find was made in the Geisental lignite deposits in Germany, where a mixture of plants, insects and other animals from all climatic areas of the world were found buried together in one common grave. In some cases leaves have been deposited and preserved in a fresh condition, the chlorophyl being still green, so that the 'green layer' is used as a marker during excavations. Among the insects are beautifully coloured tropical beetles, with soft parts of the body, including the contents of the intestines, preserved intact. Under normal conditions such materials decay or change in colour within a few hours of death, so that preservation by inclusion in an anaerobic and aseptic medium must have been sudden and complete. The same terrifying event that destroyed these inoffensive life forms also disposed of the plesiosaurs, the mesosaurs, and the other great marine reptiles. The turbulent seas were too wild, too hostile, for them to survive. Unable to escape, they joined the fate of their terrestrial brothers. Also caught in this death trap were the gigantic winged reptiles, the pterosaurs, with their 20-foot wingspan; the sky, too, made its sacrificial contribution. Only a worldwide, all-encompassing catastrophe can account for this phenomenon. There is simply no other way to explain the death of these individual species. Death took all of them – and buried them in the deep recesses of nearly every continent on the earth.

It was an end that included every living thing, but it was not until recently that scientists embarked on experiments which attempted to simulate the rapid burial process that seems to have been involved in the preservation of the telltale remains. If that were not feasible, they hoped at least to discover what natural method was responsible for the instant preservation.

Dr Coffin described some of the effects in the experiments conducted by the scientists Zangerl and Richardson. He explained:

'In attempting to evaluate the rate of fossilization in the Pennsylvania black shales of Indiana, they placed dead fish in the protecting wire cages and dropped them into the black muds at the bottom of several Louisiana lagoons or bayous. These black muds are thought to resemble the sediments from which the dark shales were derived. To the great surprise of the investigators, fish weighing from one-half to three-fourths of a pound were found to have all the soft parts reduced and all the bones completely disarticulated in six and one-half days! Decomposition to the state of total disarticulation apparently occurs at great speed, perhaps in less time than indicated above, since none were checked before six and one-half days. Delicate fossil fish showing every minute ray and bone in position are common and must represent a burial by oxygen- and bacteria-excluding sediments within hours of death if this experiment is a valid indication.'[15] The fossil remains that we find today can only be the result of burial by a sudden convulsive and overpowering flood.

But there is more to the fossil world than the bones of the giants and the delicate structures of the fish. Fascinating discoveries have emerged through conscientious study of the coal deposits that lie under much of the world's visible surface. In attempting to evaluate the various factors that led to the formation of the coal beds, Dr Coffin has uncovered some interesting facts. Limiting his comments to mineable coal seams, he points out that 'the thickness of peat needed to produce one foot of coal depends on a number of factors, such as the type of peat, the amount of water in the vegetable matter, and the type of coal. The scientific literature on coal gives figures ranging from a few feet to as many as twenty. Let us assume that ten feet would be near the average figure. On this basis, a coal seam thirty feet thick would require the compression of 300 feet of peat. A 400-foot seam of coal would be the result of a fantastic 4000 feet of peat . . .

'There are few peat bogs, marshes, or swamps anywhere in the world today that reach 100 feet. Most of them are less than 50 feet. A much more reasonable alternative theory is that the vegetable matter has been concentrated and collected into an area by some force, undoubtedly water . . .

'The concept of a global deluge that eroded out the forests and plant cover of the pre-Flood world, collected it in great mats of drifting debris, and eventually dropped it on the emerging land or on the sea bottom is the most reasonable answer to this problem of the great extent and uniform thickness of coal beds.'[16]

The big question is, *When* did this happen? The geological age theory, based on slow accumulation, obviously does not have the answer. But if we cannot account for the various phenomena as part of the geological age theory, which estimates its time periods in the millions of years, then there must be other signposts that will direct us to a time slot that is more logical and more realistic.

Archaeology may have the answer – or at least credible indications leading to an acceptable answer.

It is certain that we will never be able to determine an exact date for the great destruction, since no records have been found thus far to substantiate a historical date earlier than 3500 BC. All other datings – some claiming that remnants of civilizations centred around Jericho should be set at approximately 5000 BC – become rather nebulous when exposed to scientific critique, for as yet there is no foolproof dating method. Perhaps the answer lies in ascertaining the dates for the rise of the world's great ancient cultures. The Sumerians have given us the oldest known historical texts, but even though these inscriptions take us back to 3000–3500 BC, the origin of the authors remains a mystery. They came from somewhere, but that's all the archaeologists will agree on. Dr Samuel Noah Kramer, research professor of Assyriology at the University of Pennsylvania, says:

'The dates of Sumer's early history have always been surrounded with uncertainty, and they have not been satisfactorily settled by tests with the new method of radiocarbon dating. Be that as it may be, it seems that the people called the Sumerians did not arrive in the region until nearly 3000 BC.'[17]

Egyptian history doesn't have the answer, either: 'We think that the first dynasty began not before 3400 and not much later than 3200 BC,' says H. R. Hall, famed Egyptologist. 'A. Scharff, however, would bring the date down to about 3000 BC, and it must be admitted that his arguments are good and that at any rate it is more probable that the date of the First Dynasty is later than 3400 BC than earlier.'[18]

The Chinese have set 2250 BC as the beginning date for their history, while traditional Bible chronology estimates 2448 BC as the year for the Flood occurrence. All dates are approximate, as they have been derived by archaeologists with varying backgrounds. True, the modern carbon-dating methods have been a notable help when probing into the dust of the ages, but all the so-called final dates are still the result of assumptions and assertions, especially when we approach the 3500–3000 BC mark. The civilizations we encounter prior to those years present considerable mystery as to their origin. I recall from my studies in Egyptology that even the origin of the Egyptian *Book of the Dead* remains obscure. E. A. Wallis Budge comments in *The Book of the Dead: Papyrus of Ani:*

'The evidence derived from the enormous mass of new material which we owe to the all-important discoveries of the mastabah tombs and pyramids by M. Maspero, and to his publications of the early religious texts, proves beyond all doubt that all the essential texts comprised in the *Book of the Dead* are, in one form or another, far older than the period of Mena (Menes), the first historical king of Egypt. Certain sections, indeed, appear to belong to the predynastic period.

'The earliest texts bear within themselves proof not only of having been composed, but also of having been revised, or edited, long before the copies known to us were made, and, judging from many passages in the copies inscribed in hieroglyphics upon the pyramids of Unas (the last king of the Fifth Dynasty, about 3333 BC), and Teta, Pepi I, Mehti-em-sa-t, the Pepi II (kings of the Sixth Dynasty, about 3300–3166 BC), it would seem *even at that remote date, the scribes were perplexed and hardly understood the texts which they had before them.* [Italics are mine. R.N.][19] To fix a chronological limit for the arts and civilization of Egypt,' he concludes, 'is absolutely impossible.'[20]

Can you imagine the scribes of 3300 BC copying funeral texts and not knowing their meaning or their origin? What's more, they had to copy what was left to them from a historical period which in their day was already a nebulous recollection.

On returning to the United States from my second Noah's Ark Expedition in late 1960, one of our team members, Professor Arthur J. Brandenberger, professor of photogrammetry at Ohio State University, received a curious letter from George F. Dodwell, retired government astronomer and director of the Adelaide Observatory in South Australia. Fascinated by our expedition into eastern Turkey to find the remains of the legendary ark on Mount Ararat, he shared with us an intriguing bit of information. He wrote:

'I am especially interested in such a remarkable result, because I have been making during the last twenty-six years an extensive investigation of what we know in astronomy as the secular variation of the obliquity of the ecliptic, and from a story of the available ancient observations of the position of the sun at the solstices during the last three thousand years, I find a curve which, after allowing for all known changes, shows a typical exponential curve of recovery of the earth's axis after a sudden change from a former nearly vertical position to an inclination of 26½ degrees, from which it was returned to an equilibrium at the present inclination of 23½ degrees, during the interval of the succeeding 3194 years to AD 1850.

'The date of the change in the earth's axis, 2345 BC,' he continued, 'is none other than that of the flood recorded in the Bible, and the resulting conclusion is the Biblical account of the flood as a universal one, together with its story of Noah's Ark, is historically true.'[21]

Is it probable that the few human survivors of the Flood, whether the leader was called Noah, Nu-u or Nu-wah, depending on the tradition, were not so primitive after all? Can we assume that they carried with them sufficient knowledge of the antediluvian era to enable them to give a rapid start to the new civilizations that sprang up 'out of nowhere' in the years immediately following the Flood?

There have been many discussions recently to determine the size of the antediluvian population, the results of which can only be speculative; however, the facts must be presented in order for us to understand the size of the pre-Flood race. Regardless of the profusion of historical and legendary indications of a global flood, there are many scientists who steadfastly believe that the Biblical account, the Chaldean account and other Deluge stories merely depict local inundations. They maintain that humanity was still in its infant stage and emotionally bound to the Middle East, the place of its birth. To these primitives, these scientists reason, it seemed as if what in reality was only a local flood had indeed swallowed up the entire world. Their communications were so rudimentary, their understanding so limited, that to them nothing else existed but Mesopotamia. It is a simple theory and certainly does not provide an explanation for the worldwide distribution of Flood traditions. Everywhere our planet continues to cough up bits and pieces of information that point towards an all-encompassing catastrophe, but civilizations totally washed away by devastating waves do not tell a tale, except a silent one, and even the billions of fossils and artifacts that are daily uncovered relate only a part of the complete story.

To reconstruct the events that happened before a flood destroyed all of mankind cannot be done without combining archaeology, geology, conventional history and Biblical history into one account. Only a scrutiny of the Biblical story of the Flood, coupled with the findings of twentieth-century biologists and experts on population growth, can furnish us with possible guidelines on which to build our hypothesis.

It is in the fifth chapter of the book of Genesis that we find the statistics in connection with the age of the antediluvian patriarchs. According to this account, Adam, considered to be the first man, lived 930 years; Seth, the second patriarch, 912 years; Enos, next in line, lived 905; Cainan, 910; Mahalaleel, 895; Jared, 962; Enoch, 365; Methuselah, 969; Lamech, the ninth patriarch, lived 777; while Noah, the hero who escaped across the treacherous waters of the Flood, reached an age of 950 years.

Fantastic? Yes, it is, when compared to our present lifespan; however, there are scientists who readily accept it as reality. Dr Hans Selye, Director of the Institute of Experimental Surgery of Montreal University, says:

'Medicine has assembled a fund of knowledge that will now

serve, I believe, as a point of departure for studying the causes of old age. If the causes of ageing can be found, there is no good medical reason to believe that it will not be possible for science to find some practical way of slowing the process down, or even bringing it to a standstill.'[22]

There are numerous factors which may have contributed to the longevity of the early inhabitants of this planet. Originally man was undoubtedly endowed with considerably more vitality than we have today, or he would not have been victorious in the struggle for survival. Scientific discoveries lead us to believe that the pre-Flood world had a more agreeable climate, and areas that now are deserts once blossomed, and tropical flora grew in the polar regions. Plants thrived on rich virgin soil, but the seething waters of the Flood churned up the top layer of the earth's crust, and when the Deluge ended and the water receded into the gigantic basins now known as oceans, it tore the fertile layer of soil from the surface and washed it down to the ocean floor, leaving the surviving members of the human race with only a vestige of the nutrients available before the Deluge. For the remaining few, the earth was no longer the planet they once had known. Everything had changed, including the basic source of nutrients needed for human survival.

Today, conditions are even worse. Our degenerate race now lives on vitamin pills, and a majority of the earth's population goes to bed at night with empty stomachs. Five children per family is still considered an acceptable average if taken on a worldwide basis, considering that a woman's reproductive capacity is limited to approximately thirty to thirty-five years of her life. Not too long ago, eight, ten or even twelve children were the accepted norm in many countries. It is rather obvious what the longer lifespans of the antediluvians could have done to their population expansion. Assuming that the reproductive capacity of the pre-Flood mother was half her age – as it is for women today – then we come to the sobering conclusion that 400 years of childbearing per woman may not have been so unusual.

Allowing the average antediluvian mother this reproductive capability, we must conclude that a family of eighteen or twenty children is not an unreasonable assumption. Using the genealogy mentioned in the book of Genesis and accepting that there were ten generations from Adam to Noah, the development of the antediluvian population may have been like this:

First generation	2
Second generation	18
Third generation	162
Fourth generation	1458
Fifth generation	13,122
Sixth generation	118,098
Seventh generation	1,062,882
Eighth generation	9,565,938
Ninth generation	86,093,442
Tenth generation	774,840,978

Add to the last figure an estimate for the surviving members of previous generations, and the number 900,000,000 doesn't seem so unrealistic – a figure close to the world's population 150 years ago. How fast population figures can change became evident in a report published in 1959, indirectly supporting the premise of a large antediluvian population.

'During the first half of the nineteenth century,' the report states, 'world population figures reached 1 billion; in 1930 the figure was about 2 billion. In 1957 and 1958 alone the earth's population increased by 90 million, a figure twice the population of France, and the world is expected to have 3 billion inhabitants by 1962. The acceleration of population growth in underdeveloped countries is especially spectacular. Annual increases of 2 per cent or more are usual in most of these countries, and in some there is a growth of 3 per cent.'[23]

Interesting statistics, but old ones, for today's population is already in excess of 4 billion, *an increase of 2 billion since 1930.* Thus, assuming high birth rates and low death rates to have been the case in the pre-Flood period, the strong possibility exists that the antediluvian world was populated by a race of people

---

PATRIARCHAL AGES AT MATURITY AND DEATH

Graph showing the sudden decrease in the age of the patriarchs [at maturity and at death] after the Deluge. The horizontal "generation" scale is not intended to be an accurate time representation but an arbitrary assigned "generation number". It is not certain that the genealogies recorded in Genesis are consecutive, and of course the generations vary in length. [Adapted from William R. Vis, "Medical Science and the Bible", *Modern Science and Christian Faith; p. 241.*]

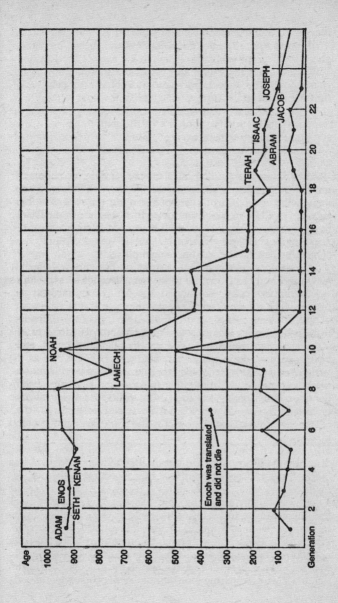

that covered practically the entire globe!

In *Medical Science and the Bible*, William R. Vis prepared a graph indicating the contrast between the ages of the patriarchs before and after the Flood and points out that 'a study of this chart shows in a striking way that something extremely significant happened to the earth and to man at the time of the Flood. It would seem that whatever this was, it probably removed the dominant factor for the long life of the patriarchs . . . Could some antediluvian climatic or other condition have been extremely favourable for long life in man? Perhaps future scientific research will cast some light on this.'[24]

After having examined all possible angles of population increase before the great catastrophe, John C. Whitcomb comments, 'We are confident, therefore, that our estimate of a population of one billion people on the earth at the time of the Deluge is very conservative; it could well have been far more than this. A population of this order of magnitude would certainly have spread far beyond the Mesopotamian plains.'[25]

The memory of this global destruction has survived roughly 5000 years and has given rise to the many legends and traditions we encounter today. Noah and his family undoubtedly are responsible for the multiplicity of existing legends. Their memories of antediluvian events were coloured by their own individual experiences, interpretations and personal recollections of the same terrifying experience. The minute changes in each story developed with the passing years into major differences, as the various versions stayed alive through successive generations, being told and retold from father to son – and when the descendants of Noah finally left the Ararat region to travel into the hinterland, they took with them their developing history.

Whatever antediluvian knowledge did come through the Flood intact must have been preserved by the survivors about the lone ark. But even though the voyage on the vessel saved the seed of humanity, it probably had a restricting influence on what was taken along. A special comment in connection with this was made by Arnold J. Toynbee in his *A Study of History*, where he notes that in migrations by ship the everyday possessions of the migrant have to be dismantled and packed for the voyage to the new land, and unpacked and reassembled upon arrival. Anything that cannot withstand the ocean voyage is usually abandoned, and many things which the migrant does manage to take with him may never be assembled in the original form again.

Most legends which mention the number of people who survived

32

the Flood speak of no more than eight, and if this is true, we can be certain that it would have been impossible for this small group of bewildered survivors to reproduce all of the aspects of the technological environment that were known to the antediluvian civilization. A technological society is dependent of course upon a large population for its maintenance; thus, not only were there not people immediately following the Flood, but also the technical resources were gone, as were the availability of natural resources and the co-ordination, specialization and integration of labour by both man and machine. Consequently the degree of civilization that the surviving human family was able to re-establish in the post-Flood era was severely reduced, and they probably preserved only those elements of technology which were useful in supplementing their limited resources on the new land. There certainly was no shortage of knowledge in the antediluvian world. Of particular interest in regard to this are the recurring traditions telling us that before the world cataclysm there existed an era of superior knowledge. The Greeks, who inherited the lore of the Middle Eastern civilizations, were well aware of the possibility of even earlier civilizations that had been destroyed through natural causes. Philo of Alexandria (c. 30 BC–40 AD) wrote, 'By reason of the constant and repeated destructions by water and fire, the later generations did not receive from the former the memory of the order and sequence of events.' Plato, in the *Timaeus*, recorded what the Egyptian priests had told his ancestor, Solon: 'There have been and there will be again many destructions of mankind,' and when civilization is destroyed, 'you have to begin all over again as children.'

The Sumerians, who are thought to have founded the first civilization after the Flood (Gen. 10:8–12) also recognized a pre-Flood civilization antedating their own. In 1922, the Weld-Blundell Expedition, excavating at Larsa – a few miles north of Abraham's city of Ur – discovered what is now called the Weld Prism, safeguarded in the Ashmolean Museum at Oxford. The Prism contains a history written by a scribe named Nur-Ninsubur in approximately 2100 BC. In his account he records the list of ten pre-Flood kings and ends his writing with the sad words, 'and the Flood overthrew the land.' The Sumerians and later the Babylonians and Assyrians further recognized the pre-Flood era as the source of superior literature. One Babylonian king recorded that 'he loved to read the writings of the age before the Flood.' Assurbanipal, who founded the great library of Nineveh, also referred to the great 'inscriptions of the time before the Flood.' It must be remembered that the reason the pre- and post-Flood civilizations advanced at a more rapid rate than our own today is that they were using their brains to their fullest capacity, whereas we, according to the physiologists, are using only one-sixth of ours.

In Guatemala, the ancient Mayas recorded in their sacred book, the *Popol Vuh*, that the 'first men' possessed tremendous knowledge: 'They were able to know all, and they examined the four corners, the four points of the arch of the sky and the round face of the earth.'[26] The ancient Chinese also recognized that before them were 'giants': 'Men twice as tall as us,' who once inhabited the 'realm of delight' but lost it by not living 'by laws of virtue.'[27]

Many fragments of technological know-how of unknown origin have surfaced over the years and have usually been ignored by modern science, because of the mystery they represent.

There is a fact that cannot be ignored and must of necessity be considered when evaluating new discoveries. In the making of our own civilization, we are aware that science has always advanced by means of the scientific method. Through this method, long observation and much thought concerning the repeated processes observed have led to scientific discovery. At the same time technology has taken what science has discovered and applied or transformed it into a usable form. Science is the discovery of natural processes; technology is the application of the discovery for practical purposes.

Both science and technology have progressed not through chance, but rather through increasing degrees of sophistication. In science, each discovery has become the basis for research leading to new discoveries; in the same way, technology has taken known scientific principles and applied them towards new developments, which in time have been added to others and have led to more sophisticated developments. In other words, each level of sophistication has been dependent on the previous levels. This progression by which fundamentals are combined to create a total effect greater than the separate effects of the component fundamentals is called synergism.

In the era of technological development with which we are the most familiar, our own, synergism and not blind chance has been the governing factor. The first technologists of western civilization did not begin by trying everything possible and stumbling on their developments by accident. Instead there was a necessary connection between the ideas and analogies with which

34

scientific discovery presented them and the results they were striving towards.

If technological development by synergism was characteristic of our civilization – and I am sure no one can question this – then it is difficult to imagine that it would not have been the same in the technological development of any earlier advanced civilization. We find in the Genesis account and in the historical record that the antediluvians did in fact possess the necessary potential by which they could have advanced by synergism to develop a technology at least as sophisticated as our own.

Archaeology tells us nothing at all about the pre-Deluge period, for most archaeologists completely disregard an antediluvian era in history, as they theorize that the Flood was merely a local event and not worth considering in the scheme of world history. Thus, the only way we can attempt to reconstruct this extremely important period is by examining the historical and genealogical records mentioned in the 'third source' of historical knowledge – the Bible.

Regardless of the criticism heaped on the Bible, there is a serious trend among archaeologists to regard its historical books as reliable.

Until the eighteenth century, few scholars doubted the trustworthiness of the Bible as a historical book of antiquity. The Creation, the story of the Flood, the sojourn of the Israelites in the desert were all deemed facts. But then things changed. The infamous 'Age of Reason' breached the walls of faith, and with the coming of the new nineteenth-century theories of evolution and materialism, the historical Bible account was relegated to mythology, and the 'enlightened age' began to regard the Bible as a well-contrived compilation of fables. Harry M. Orlinsky comments in *Ancient Israel*, 'The heroic doings of the patriarchs, Abraham, Isaac, and Jacob, as described in the Book of Genesis, were discounted as mere myth. The very existence of Moses was doubted. Joshua was believed to have had little or nothing to do with the Israelite conquest of Canaan. David and Solomon were considered greatly overrated . . .' This negative attitude to the Bible was reflected in more recent times, for example, in the writings of the well-known philosopher Bertrand Russell and the historiographer R. G. Collingwood. 'For today, in considerable degree,' Orlinsky continues, 'the pendulum has swung the other way. Modern historians do not, to be sure, accept every part of the Bible equally as literal fact. *Yet they have come to accept much of the Biblical data as constituting unusually reliable historical documents of antiquity*, documents which take on new meaning and pertinence when they are analyzed in the light of the newly discovered extra-Biblical sources. Indeed, even the mythical parts of the Bible are now generally regarded as reliable reflection of fact, empirically grounded, and logical in their way . . . More and more the older view that the Biblical data were suspect and even likely to be false, unless corroborated by extra-Biblical facts, is giving way to one which holds that, by and large, the Biblical accounts are more likely to be true than false, unless clear-cut evidence from sources outside the Bible demonstrates the reverse.'[28]

Professor William F. Albright, the world-renowned archaeologist of Johns Hopkins University in Baltimore, adds, 'The Bible now forms, humanly considered, part of a great whole, to the outside parts of which it can be related. Its languages, the life and customs of its peoples, its history, and its ethical and religious ideas are all illustrated in innumerable ways by archaeological discovery.'[29]

With these endorsements of the Bible as a historical document, we dare to venture into the first chapters of Genesis in order to help us trace the development of the human race and the technological and cultural achievements with which it coincided. There is nothing else that can penetrate the history of the pre-Flood period. Even modern dating methods are of no help, for they are applicable only to artifacts, not accounts, and these dating techniques cannot take us back further than 5000 years. One of the world's foremost experts in modern dating methods, Dr W. F. Libby, who won the Nobel prize for his research on carbon-14 dating, was astonished to find that this was so. Trusting that his new-found tool would enable science to pull back the curtain of time, he was shocked to be confronted with science's limitations.

'You read statements in books that such and such society or archaeological site is 20,000 years old,' he commented. 'We learned rather abruptly that these numbers, these ancient ages, are not known accurately; in fact, it is at about the time of the First Dynasty of Egypt that the first historical date of any real certainty has been established.'[30]

Genesis tells us that Cain and Seth were born in 4969 BC and 4924 BC respectively, and upon maturity became the progenitors of two distinct races, isolated from one another. Known as Cainites and Sethites, they developed totally different life-styles. A close examination of the names and identifications given to the various tribal leaders or patriarchs that ruled the two races tends to bring us to a better understanding of their capabilities – and the conclusions we reach clash violently with the concept of crude cave men dragging their wives by the hair on the road to marital bliss.

We know from the record that Cain lived in the land called Nod, meaning 'place of exile'. Genesis mentions that the land where he settled did not 'yield her strength'. There is no further information concerning the early years except that in approximately 4784 BC a son, Enoch, was born. Sometime between this date and Cain's death in 4059 BC, Cain gathered his descendants together and built a city called Enoch City, named after his first son. Several observations can be made regarding the construction of this first city, which would require the development of a high order of mathematics; the manipulation of building materials, seemingly stones and wood; and a knowledge of architecture. A city also presupposes the beginning of some form of social and political organization, not only for the construction phase, but also for its subsequent maintenance and expansion. Upon assembling his descendants into one dwelling place, Cain set himself up as the first ruler over men, and by naming the city after his eldest son, he moved to establish a dynasty of rulers that would perpetuate his name.

Enoch was the next Cainite, supposedly born in 4784 BC. Little is known of the personal histories of the descendants of Cain other than their names and sequence of descent as recorded in Genesis, chapter four. However, it was the ancient practice to give men names that commemorated their status or character or major events in their lives. From their names we can now perceive something about each of the Cainites. The name Enoch means 'devoted, the initiated (into secret learning), a teacher,' and the implication is that Enoch was a man of knowledge, particularly mystic knowledge. The fact that Cain named his city after Enoch suggests that the city was to be not just a political centre but a religious one as well. Enoch City, with Enoch as its high priest, may have possessed its own sanctuary and sacrificial system.

Irad, the son of Enoch, was born in approximately 4599 BC and died about 3689 BC. His name means 'a townsman, a prince of a city'. The city where Irad was ruler was of course Enoch City, and its throne was his inheritance from his grandfather, who died during Irad's lifetime. This indicates that the dynasty of Cain remained intact at least to the third generation.

Mehujael, the son of Irad, entered the scene at 4415 BC, his name meaning 'smitten of God'. We do not know the manner in which this Cainite was struck down, whether by disease, malformation or natural disaster, but the record indicates that his condition was regarded as a punishment.

Methusael, the next in line, was born in 4367 BC and died about 3398 BC and, as his name indicates, was 'a great man before God'.

Lamech, the son of Methusael, was born in approximately 4180 BC and died about 3403 BC. While the history woven around his father seems to indicate that there may have been a mixing of the two races for a while, perhaps for a specific but unnamed purpose, it doesn't seem to have affected the development of the Cainites and Sethites as separate races. However, immediately after the races had intermingled, the account indicates that open lawlessness was common in this developing society. Lamech means 'a strong young man, a hero'. The record indicates that he was not only strong, but also a murderer and the first polygamist mentioned. His declaration of the murder is lyric in form. It was at this time that the arts emerged. Lamech's first wife, Adah, was an artist, her name meaning 'ornament, decoration, elegant'; the name of Zillah, his second wife, meant 'a shadow figure, a maker of sound, a player', and she may well have been the first actress. Other members of the family are described as 'such as dwell in tents', 'such as have cattle', 'wanderer', or 'adventurer'. No doubt the Cainites had become quite mobile and were roaming the countryside, seeking land for their flocks, and exploring various forms of the arts. Was it perhaps the loneliness in the fields that inspired the shepherds to try out new ways of accompanying their singing? We are not certain of this, yet Jubal, Adah's second son, was 'the father of all such as handle the harp and the organ,' indicating the beginnings of the fine art of instrumental music, both string and wind.

With the arrival of Tubalcain (3860 BC), son of Zillah, technology entered the antediluvian world. Genesis 4:22 calls him 'an instructor of every artificer in brass and iron'. The production of metals is, of course, a very significant step in the development of a civilization, for it facilitates the way to higher forms of technology and more sophisticated tools and machinery. Iron presupposes knowledge of the techniques of ore extraction and purification; brass presumes knowledge of copper and zinc and their combinations in the production of metallurgical alloys. An artificer of metals is one who hammers, cuts and polishes metals; the shaping of metal instruments thus is implied. The form that these instruments took can be taken from the name Tubalcain itself. It means 'the brass of Cain, brass weapons, a weaponsmith'. Tubalcain therefore was the manufacturer of the first known metal weapons in history, and the production of weapons, of course, suggests war, or at least the threat of war.

In 1968, Dr Koriut Megurchian of the Soviet Union unearthed what is considered to be the oldest large-scale metallurgical factory in the world, at Medzamor, in Soviet Armenia. Here, 4500 years ago, an unknown prehistoric people worked with over 200 furnaces, producing an assortment of vases, knives, spear-heads, rings, bracelets, etc. The Medzamor craftsmen wore mouth-filters and gloves while they laboured and fashioned their wares of copper, lead, zinc, iron, gold, tin, manganese and fourteen kinds of bronze. The smelters also produced an assortment of metallic paints, ceramics and glass. But the most out-of-place discovery was several pairs of tweezers made of steel, taken from layers dating back before the first millennium BC. The steel was later found to be of exceptionally high grade, and the discovery was verified by scientific organizations in the Soviet Union, the United States, Britain, France and Germany.

French journalist Jean Vidal, reporting in *Science et vie* of July 1969, expressed the belief that these finds point to an un-known period of technological development. 'Medzamor,' he wrote, 'was founded by the wise men of earlier civilizations. They possessed knowledge they had acquired during a remote age unknown to us that deserves to be called scientific and industrial.'

What makes the Medzamor metallurgical site interesting to us is that it is within fifteen miles of Mount Ararat – the landing site of the survivors of the destroyed antediluvian civiliz-ations.[31,32,33]

The development of the Sethites was in no way behind that of the Cainites. Seth, born about 4924 BC, is not known as either a city dweller or a city builder, but seemed to have lived quietly on the fertile soil provided by the four rivers mentioned in the early records. Yet something happened to undermine the physical well-being of the Sethites, for his son, born in approxi-mately 4819 BC, carried the name Enos, which means 'mortal, weak mankind'. This may be indicative of disease taking its toll. The Hebrew Aggadah comments that during Enos's lifetime men's faces became more apelike.

But there are more men to be considered in this short list of antediluvian greats. Cainan (4729 BC) was 'an industrious man, a craftsman' as his name indicates, and he may have been the first to usher in the development of sophisticated tools – years before Tubalcain embarked on his weapons production. Simple tools had already been replaced by more complex ones which were perhaps used in carpentry, pottery, weaving, masonry, etc. By now the Sethite population was probably large enough so that its working force had become diversified, and as a craftsman Cainan typified the growing specialization of labour that has always accompanied an expanding culture.

There are many others who should be listed here, but their reported professions were more spiritual, and because we are more concerned with the technological aspects of pre-Flood life, we will single out only three more Sethites: Methuselah (4367 BC); Lamech (4180 BC), son of Methuselah; and Noah (3998 BC), son of Lamech.

With Methuselah we enter an era of open warfare. Bearing a name meaning 'man of the flying dart, man of the arrow, man of war', he undoubtedly was a military man – and a successful one, for he lived longer (969 years) than any of the other patriarchs. His adversaries may have been indicated already when we discussed the descendants of Cainan. Methuselah, living in the same period as Tubalcain, the weaponsmith, was a master of several weapons, including the arrow and the 'flying dart'. Could this possibly have been a missile or a rocket, and is this the first indication of a major armed conflict between the two civilizations, the Cainites and the Sethites? We know from historical accounts that Jubal, Tubalcain's half brother, was the first to spread his influence by venturing into new lands. Perhaps his men threatened to invade the territories already inhabited by the Sethites.

Lamech probably followed the profession of his father Methuselah, but he died at the age of 777, which was young compared to the average lifespan of 912 years for the pre-Flood patriarchs, not counting Enoch. He was even outlived by his father. Could his death have been the result of wounds inflicted in battle?

Biblical Noah, who was born in 3998 BC and died in 3048 BC, was the last of the antediluvian giants. Mentioned in mythology by various names such as Nu-u, Nu-wah and others, he was the man who, with his wife, his three sons and their wives, plus a representation of the animal kingdom, braved the turbulent waves of the Deluge in their vessel, thus ending the reign of the ten patriarchs. His sons, Shem, Ham and Japheth, comprised the eleventh generation, and since part of their lives were lived after the Flood, they must be considered post-Flood patriarchs.

We will never know the exact extent of the knowledge that was lost. Noah and his family were capable of reproducing only that of which they had personal knowledge, and this of course was limited because the only ones to reach the previously attained high standard of development were Noah and his sons. The technology that survived the Flood was the product of possibly ten generations of synergistic development, the intermediary steps of which had been totally destroyed. Noah and his family at least had the memory of a technological environment to fall back on, but Noah's sons' children did not. All they knew and understood was the reduced civilization of their parents, a post-Deluge civilization that did not possess in its structure the gradual development behind the technological elements their parents preserved. In time the children began to lose the knowledge of the fundamental principles; and when the technological elements broke down and could not be replaced, they were discarded.

The one chance to unify their descendants into an orderly, well-organized society came at Babel, a city mentioned in the Book of Genesis. The story relates that at the Tower of Babel – the first skyscraper in history – a serious attempt was made to structure the rapidly growing population under one centralized authority, but this plan failed because their language became confused, and the communication necessary for the re-creation of the super-civilization that had existed before the Flood was destroyed.

In Chapters 10 and 11 of Genesis we find two other factors that hampered a complete restoration of antediluvian technology. The descendants of Noah became divided into nations and races, with the result that the common background they had shared was lost, with some people preserving the knowledge and others losing it. For those who managed to retain some pre-Flood knowledge, especially the more advanced elements of technology, the national and racial divisions of mankind caused a downward trend. Hindu records actually speak of highly destructive war waged with nuclear weapons, and there is every indication that these clashes occurred relatively soon after the Flood. Since very little survives a total nuclear war, it is conceivable that several advanced civilized centres vanished simultaneously, another probable factor in the disappearance of antediluvian technology in the post-Flood era.

Genesis 11 brings in yet another element. The generation following Noah suddenly shows a decided decrease in the average lifespan, from more than 900 years to approximately 100 years. This, of course, severely limited the individual's chance to acquire knowledge and experience, and with shortened lives, the generations passed more rapidly. Considering that there were now more human memories involved in relaying and passing on all information, the likelihood increased that facts would become distorted.

Writing, an element we consider extremely vital if not indispensable to our society, seems to have been completely unknown to the ancients. Is it possible they did not need it? Puzzling though it may be, there are indications that this may have been the case, because writing was actually considered to be a step backward for civilization, rather than a step forward.

In his *Phaedrus* the Greek philosopher Plato wrote about the legend of Toth, the Egyptian god who supposedly discovered the use of letters. In order to boast of his invention, the god demonstrated their use to King Thamus and claimed that this newfound form of communication would be an aid to wisdom. But the king condemned Toth and told him that just the opposite would be true. Writing, he judged, would encourage forgetfulness in the minds of those who learned, because they would not cultivate their memory. Students would learn the appearance instead of the reality of wisdom, reading and repeating words without knowing their meaning. He declared that writing would limit knowledge, not expand it. And it happened as he prophesied, for as we have already seen, the early Egyptian funeral texts are examples of just that development, as the scribes copied the texts without knowing their significance. Many historians have noted that the secularization of the written word in the past was not always an indication of the rise of a civilization; sometimes it was an omen of its decline.

But regardless of King Thamus's objections, writing did become a reality and wielded its influence on the growing societies. It was a help to some, but acquiring the knowledge to write brought with it serious limitations. Memory was no longer *the* factor to rely on; now the storehouses of words became the prime factors in the dissemination of knowledge. Whereas before the Flood, technological facts had been transmitted from father to son, from scientist to scientist, now huge depositories of written knowledge were substituted for oral tradition, and as a result a privileged few became the sole custodians of this accumulating knowledge. When the rampaging armies of expanding nations invaded the lands, the great libraries of the world became the innocent victims of wanton destruction. Many of history's missing pages were torn out in those calamitous years. 'The famous collection of Pisastratus [Pisander] in Athens (sixth century BC) was ravaged. Fortunately the poems of Homer somehow survived. The papyri of the library of the Temple of Ptah in Memphis were totally destroyed. The same fate befell 200,000 volumes in the library of Pergamus in Asia Minor. The city of Carthage, razed by the Romans in a seventeen-day fire in 146 BC, is said to have possessed a library of half a million volumes. But the greatest blow to history was the burning of the Alexandrian library in the Egyptian campaign of Julius Caesar, during which 700,000 priceless scrolls were irretrievably lost . . . There was a complete catalogue of authors in 120 volumes with a brief biography of each author.'[34] The library of Alexandria, however, survived this destruction and once again became a centre of learning, the most important book depository in the Mediterranean world, until Omar, the second Caliph of Islam, used its millions of book rolls to heat the city's bathing facilities in AD 640. For six months the fires roared, fuelled by the knowledge of the ancients. The Caliph decreed: 'The contents of these books are in conformity with the Koran or they are not. If they are, the Koran is sufficient without them; if they are not, they are pernicious. Let them therefore be destroyed.'[35]

Tomas, author of *We Are Not the First*, comments that the fate of libraries in Asia was no better, for the Emperor Ch'in Shih Huang Ti of China caused all historical books to be burned in 212 BC. Leo Isaurus sent 300,000 books to the incinerators in Constantinople in the eighth century. 'The number of manuscripts annihilated by the Inquisition . . . in the Middle Ages can hardly be estimated. Because of these tragedies we have to depend on disconnected fragments, casual passages, and meagre accounts . . . The history of science would appear totally different were the book collection of Alexandria intact today.'[36]

But all was not lost.

After the Second World War the discovery of the Dead Sea Scrolls created a sensation among Biblical scholars, for these documents, dating back to the second century BC, agreed remarkably with a Biblical manuscript known as the Masoretic Text of AD 10. Somehow the texts had survived practically intact. More recently, the discovery in 1975 of the Eblite Tablets at Tel Mardiqh in Syria caused another wave of excitement. The fact that the Dead Sea Scrolls could be dated to 200 BC was in itself startling; but finding clay tablets that were inscribed in 2300 BC blew the lid off! With texts written in Canaanite and other languages, the 15,000 tablets revealed a wealth of correspondence, including political treaties, records, laws, religious texts and historical information. 'It covers an important part of the post-Flood patriarchal period,' one scholar told me recently. 'It may well provide us with details that will shed a completely new light on the civilization that existed in their day.' And it is the post-Flood patriarchal period that concerns us.

Without question, a vast storehouse of ancient knowledge has been lost over the years, but not always destroyed. Originating in the period before the Deluge, it survived the angry waves in segments of selected knowledge, carefully preserved by the memories of the family of escapees.

The antediluvians undoubtedly possessed extraordinary mental abilities, for they passed quickly through several successive stages of development. By the second generation they practiced agriculture and animal husbandry; possessed fire and simple tools; knew mathematics, astronomy and architecture; and were organizing themselves into the first urban social system. By the fourth generation, simple tools had developed by synergism into such crafts as weaving, carpentry and masonry. Finally, by the eighth generation, we witness a burst of inventiveness with the beginning of metallurgy, the art of warfare, and the development of the arts. The Genesis account of these elements of civilization may not seem impressive, but we must remember that these discoveries are the first causations from which all subsequent technological development originated. We must also keep in mind that the Bible, even though it contains much historical information, is basically a religious book, and the Biblical authors wrote down only those historical events that pertained to the development or decline of their religion. We should not expect the book of Genesis to delve into the technological aspects of the maturing civilizations. Caught up as we are in the age of development, we often fail to realize that development is a lesser achievement than causation. Development is the maturation of causation, but a causation is a distinct departure from one mode of existence to another – a completely unprecedented transformation. For example, the step forward which Tubalcain took from nonmetal to metal production must be considered a far greater accomplishment than any later developments in metallurgy. The first men to walk the face of the earth, in the first eight generations of the antediluvian era, advanced from no culture to culture by their own efforts, with no precedence whatsoever. They were the originators, the discoverers and inventors, not only of their own civilization but – through the survivors of the Flood – of all the civilizations that followed them in the post-Deluge era.

For the reason previously given, the Genesis account unfortunately furnishes us few clues to the development of the antediluvian civilization after the time of Tubalcain. However, looking at what they possessed by that time, the potential for continued advancement towards the development of a high technological civilization most certainly existed. First, they had the necessary foundation in knowledge. By mastering mathematics, metallurgy and the fine arts, the antediluvians demonstrated their analytical, inventive and imaginative capabilities. The first two fields – mathematics and metals – are necessary for the introduction of machinery, the next synergistic step in tool development. Second, they had the needed human resources. Any technology depends on a work force that is large, organized and diversified. Genesis 6 tells us that 'men 'began to multiply upon the face of the earth'. The genealogy given in Chapter 5 of the historical record clearly implies that men had large families and longer procreative periods before the Flood. Although in most cases only one son in each family is named for the purpose of tracing the line of descent, it is also recorded that each patriarch 'begat sons and daughters', implying that each of them produced at least four children. Furthermore, the age at which the patriarchs had their mentioned sons varies from 65 years for Mahalaleel and Enoch to 500 years for Noah: a range of 435 years. Thus, through the combined effects of lengthy lives and large families, the antediluvians rapidly 'filled the earth'. We know, too, that the greater portion of the pre-Flood population was organized. As early as the time of Cain, his first descendants had been gathered together into an urban society and taught to be subservient to a single political and religious head. The pre-Flood population was also diversified. From the time of Cainan, the fourth generation, the race had multiplied to the extent that they could support craftsmen and a variety of trades.

What's more, there was sufficient time for an advanced civilization to develop. From Tubalcain's discovery of metalworking in approximately 4000 BC to the year of the Flood, 3398 BC, is a period of 600 years within which further advancements could have been made. This is a vital point, for it took our civilization about 600 years to develop to what it is today – from gunpowder and printing to nuclear physics and computers. If this is where we are today after 600 years, just imagine how far the antediluvians could have advanced in the same length of time – the people who were the originators of civilization.

While our antediluvian predecessors developed a technology that was in many ways similar to our own, some of the differences were so great that historians and archaeologists are still unable to correctly identify the representative remains.

Most nonmechanically minded scholars do not realize that there are products of technology which do not resemble what we call machines – without shafts, rods or gears. For example, a network of lines traced with special metal-containing ink on specially treated paper can serve as a receiver for electromagnetic waves; a copper tube can serve as a resonator in the production of high-frequency waves; and the surface of a diamond can even be made to contain an image of the pages of 100,000 average-sized books!

The problem, however, is that, as any technology advances, its methods and forms are often simplified and may not be recognizable to a civilization of inferior knowledge and understanding. Many out-of-place artifacts (ooparts) discovered today exhibit signs of a technology that not only matches our own, but in some cases surpasses it. Some of these ooparts seem so fantastic that we simply cannot grasp their significance. We can only recognize and understand these earlier developments as we ourselves approach or reach the same stage of advancement. A disturbing question, however, is how many out-of-place artifacts have been lost or remain unidentified in the basements of modern museums, because no one knows what they are?

CHAPTER 2

Ooparts—Science in the Raw?

The discoveries of the ooparts – witnesses of our most ancient past – have thus far been accidental. They are remnants of a past we have never fully recognized or believed in. Because we do not totally comprehend the scope of this mysterious pre-Flood civilization, we stand aghast when confronted with the possible relics from such a civilization.

'But where do the cave men fit in?' is a question often encountered when one discusses the greatness of our ancestors. 'We are still climbing the ladder of social evolution,' is another frequently heard comment. 'There simply is no place for a super-civilization in our past.' And, shrugging their shoulders at such *naïveté*, the critics retreat to their niches of complacency.

Yet what are they going to do with the ooparts? Too many have surfaced over the years for them simply to be ignored.

Let's look at some of the recorded discoveries. An interesting item appeared in many of the nation's newspapers on 10 April 1967, reporting the discovery of an artifact and human remains at the Rocky Point Mine in Gulman, Colorado. At a depth of 400 feet below the surface, according to an account in the Saturday *Herald* of Iowa City, the excavators found human bone embedded in a silver vein. By geological standards, the find was estimated to be several million years old. But in addition to the bones, they uncovered a well-tempered copper arrowhead four inches long. Neither bone nor arrowhead belonged there, according to our way of thinking, yet there they were – unexplainable and certainly unexpected. The historians and geologists are unable to fit these remains into the theoretical framework of evolution; partly because of this, the find has been conveniently forgotten.

But this strange discovery is not an isolated one. In the June 1851 issue of *Scientific American* (Vol. 7, p. 298) a report concerning a metallic vase that had been dynamited out of solid rock on Meeting House Hill in Dorchester, Massachusetts, was reprinted from the *Boston Transcript*. The story said, 'On putting the two parts together it formed a bell-shaped vessel, 4½ inches high, 6½ inches at the base, 2½ inches at the top and about an eighth of an inch in thickness. The body of this vessel resembles zinc in colour, or a composition metal in which there is a considerable portion of silver. On the sides there are six figures of a flower, a bouquet, beautifully inlaid with pure silver, and around the lower part of the vessel, a vine, or wreath, inlaid also with silver. The chasing, carving and inlaying are exquisitely done by the art of some cunning craftsman. This curious and unknown vessel was blown out of the solid pudding stone, fifteen feet below the surface.'

Where did it come from?

Neither the geologists nor the archaeologists know, but the rock from which the man-made *objet d'art* was taken was estimated by them to be at least several million years old. As is the case with many puzzling discoveries, the vase was circulated from museum to museum, and then disappeared. No doubt it is gathering twentieth-century dust somewhere in a curator's dank basement . . .

Precisely forty years later, on 9 June 1891, a somewhat similar find was made by Mrs S. W. Culp of Morrisonville, Illinois. While she was shovelling coal into her kitchen stove, her attention was drawn to one lump of coal which had broken in two, revealing a gold chain of intricate workmanship. The *Morrisonville Times* of 11 June reported, 'Mrs Culp thought the chain had been dropped accidentally in the coal, but as she undertook to lift the chain up, the idea of its having been recently dropped was shown to be fallacious, for as the lump of coal broke, it separated almost in the middle, and the circular position of the chain placed the two ends near to each other; and as the lump separated, the middle of the chain became loosened while each end remained fastened to the coal . . . This is a study for the students of archaeology who love to puzzle their brains out over the geological construction of the Earth from whose ancient depth the curious are always dropping out.'

The paper's editor really didn't know how to handle this bizarre discovery, but neither did the geologists, for the coal sample was supposedly from the Carboniferous period and so was thought to be several million years old.

The Morrisonville chain was in no way unique, for another gold artifact of unknown origin was discovered in 1844 in a quarry near Rutherford Mills, England. On 22 June of that year, workmen blasting granite out of the pit suddenly came upon a gold thread eight feet below the surface, embedded in rock judged by geologists to be sixty million years old. Investigators sent by the London *Times* reported that in their opinion the thread had indeed been of artificial manufacture.

Artifacts of precious metal have not been the only objects unearthed from solid rock. The *Springfield* (Illinois) *Republican* stated in 1851 that a businessman named Hiram de Witt had brought back with him from a trip to California a piece of auriferous quartz about the size of a man's fist, and while de Witt was showing the rock to a friend, it slipped from his hand and split upon striking the floor. In the centre of the quartz, they discovered a cut-iron six-penny nail, slightly corroded but entirely straight, with a perfect head. The age of the quartz, you wonder? Scientists conclude that it is in excess of a million years!

But this wasn't the first nail discovered. Six years before this find, Sir David Brewster made a report to the British Association for the Advancement of Science which created quite a stir. A nail obviously of human manufacture had been found half-embedded in a granite block excavated from the Kindgoodie Quarry in Northern Britain. It was badly corroded, but identifiable, none the less. Once again, the granite was determined to be at least sixty million years old.

Still another out-of-place artifact, a two-inch metal screw, was discovered in a piece of feldspar unearthed in 1865 from the Abbey Mine in Treasure City, Nevada. The screw had long since oxidized, but its form, particularly the shape of its threads, could easily be seen within the feldspar. Here too this discovery played havoc with accepted scientific theories, for how the impression of a two-inch metal screw could be found in something thought to be several million years old clearly perplexed the examiners.

The Salzburg Cube

The out-of-place objects that have been found in the various rock strata not only reveal evidence of simple metal production, but also indicate that the antediluvians had the ability to shape metal by machines and that they used metal in the construction of complicated machinery.

In 1885, in the foundry of the Austrian Isador Braun of Vocklabruck, a block of coal dating from the Tertiary period was broken open. Inside was discovered a small metal cube. Fascinated by this sudden find, Braun's son took the mysterious cube to the Salzburg Museum, where it was subjected to meticulous examination by the Austrian physicist Karl Gurls.

Tests indicated the cube was composed of a steel-and-nickel alloy. It measured 2·64 by 2·64 by 1·85 inches, weighed 1·73 pounds, and had a specific gravity of 7·75. The edges of this strange cube were perfectly straight and sharp; four of its sides were flat, while the two remaining sides, opposite each other, were convex. A rather deep groove had been cut all the way around the cube about midway up its height. There was no doubt that the cube was machine-made, and it seemed to be part of a larger mechanism.

Unfortunately the cube disappeared from the Salzburg Museum in 1910, and during the bombings of the Second World War the museum's inventory files relating to the time period when the cube was on exhibit (1886–1910) were completely destroyed. However, there is still sufficient evidence to support the authenticity of the find, for an account of its discovery was published in the scientific journals *Nature* (London, 1886) and *L'Astronomie* (Paris, 1887).

The Coso Artifact

Another equally controversial find was made more recently. On 13 February 1961, three rock hunters, Mike Mikesell, Wallace Lane and Virginia Maxey, were collecting geodes about six miles north-east of Olancha, California. On this particular day, while searching in the Coso Mountains, they found a stone located near the top of a peak approximately 4300 feet above sea level and about 340 feet above the dry bed of Owens Lake. The rock-hounds mistakenly identified it as a geode, a round stone with a hollow interior lined with crystals, though it bore traces of fossil shells. The following day, when Mikesell cut the stone in half, ruining a ten-inch diamond saw in the process, he saw that it contained not crystals but rather something totally unfamiliar. Inside were the remains of some form of mechanical device. Beneath the outer layer of hardened clay, pebbles, and fossil inclusions was a hexagonal layer of an unknown substance softer than agate or jasper. This layer surrounded a three-quarter-inch-wide cylinder made of solid porcelain or ceramic, and in the centre of the cylinder the finders discovered a two-millimetre shaft of bright metal. This shaft, the rock enthusiasts discovered, was magnetic and showed no signs of oxidation. Circling the ceramic cylinder were rings of copper, and these also had not corroded.

Not knowing what to do with their unusual find, they sent the object to the Charles Ford Society, an organization specializing in examining extraordinary things. X-rays taken of the fossil-encrusted rock revealed further evidence that the content of the 'geode' was indeed some form of mechanical apparatus. The photographs indicated that the metallic shaft was corroded at one end, but the other end was affixed to a spring or helix of metal. The Coso artifact, as it is now known, is believed to be more than just a piece of machinery. The finely shaped ceramic and metallic shaft and copper components hint at some form of electrical instrument. It bears a close resemblance to a spark plug, but there are certain features – particularly the spring or helix terminal – that do not correspond to any spark plug known today. To complicate the mystery surrounding this strange little instrument, the geologists tell us that the rock in which it was found has to be at least half a million years old.

The controversy in which these finds are enveloped concerns the dates that have been assigned to them because of the strata in which they were discovered. There is no doubt that Mrs Culp did find the gold chain in Carboniferous rock, and the discovery of the gold thread in the quarry near Rutherford Mills is also a matter of record, as are Hiram de Witt's iron nail, Sir David Brewster's report, and the metal screw found in the Abbey mine. But there is one vital factor that must not be overlooked: that the dates given for their origin cannot be relied on for accuracy. No conscientious geologist will dare attach any degree of certainty to the various dates given to the different layers of the earth's crust. It is more reasonable to look at the artifacts in the light of Deluge geology, which maintains that the stratified rock is the result of soil laid down by water. This signifies that the metal objects encased in the rocks were buried during the Flood, and thus their manufacture would date from before the Deluge. The ooparts certainly set the theories of the geologists against those of the historians, for one group holds stubbornly to the million-year-age theory, and the orthodox historians definitely are unwilling to accept the authenticity of a machine-made cube in a block of coal dating back to the Tertiary period. To them the existence of a highly advanced civilization one hundred million years ago is incredible. This exaggerated time element must be rejected. Since we acknowledge that coal is a product of vegetation destroyed, compressed and buried by water, the Salzburg cube, found in the so-called Tertiary coal, must therefore date from the pre-Flood period. Because the Coso artifact was found in sedimentary rock, we must conclude that

51

this too was deposited during the great Flood. What makes these artifacts significant to us is that they reveal that the antediluvians had progressed beyond mere metal production and had obviously learned how to utilize certain forms of energy – in this case electricity – several thousand years before the reintroduction of this knowledge into our civilization.

For years now, a slow but methodical search has been conducted to find the elusive ark of Noah, the ship that bridged the gap between the antediluvian and postdiluvian civilizations. We have always thought of it as a simple wooden ship, of which we know only the approximate dimensions. I have participated in numerous discussions speculating about the possible contents of the ship, yet in all those long, thought-provoking hours, no one ever thought of Noah and his family as members of a highly civilized race. Problems of waste disposal, ventilation, air conditioning, maintenance and lighting were disposed of with a casual wave of the hand.

'Their civilization wasn't far enough advanced for a sophisticated technology,' was the usual consensus. 'Don't look for the impossible.'

Can it be that we were all wrong?

Let's take a look at the Genesis description of the Flood and the survival vessel, and focus our attention on two references. In this account we find two indications that lead us to believe that electricity may have played a vital role in the operation of the ark. One reference is found in Genesis 8:6, where the Hebrew word *challon* or 'opening' is used, referring to the window through which Noah released the birds. The other reference, however, utilizes a different word – *tsohar* – which is translated as 'window' but does not mean window or opening at all! Where it is used (twenty-two times in the Old Testament), its meaning is given as 'a brightness, a brilliance, the light of the noonday sun'. Its cognates refer to something that 'glistens, glitters or shines'. Many Jewish scholars of the traditional school identify *tsohar* as 'a light which has its origins in a shining crystal'. For centuries Hebrew tradition has described the *tsohar* as an enormous gem or pearl that Noah hung from the rafters of the ark, and which, by some power contained within itself, illuminated the entire vessel for the duration of the Flood voyage.

Noah's light source seems to have been preserved in history for hundreds of years, for we find indications that King Solomon of Israel may have used it in about 1000 BC. An ancient Jewish manuscript entitled 'The Queen of Sheba and Her Only Son

Menyelek', translated by Sir E. A. Wallis Budge, contains this statement: 'Now the House of Solomon the King was illuminated as by day, for in his wisdom he had made shining pearls which were like unto the sun, the moon and the stars in the roof of his house.'

In view of this, it is not surprising that Solomon himself once wrote, '. . . there is no new thing under the sun. Is there any thing whereof it may be said, See, this is new? it hath been already of old time which was before us.' (Ecc. 1:9–10.)

Electricity in one form or another has surfaced throughout the centuries. According to the historian Josephus Goriondes, Alexander the Great wrote to his teacher during his conquest of Persia that an island located off the coast of India was inhabited by men who ate raw fish and spoke a language akin to Greek. They believed that at one time Cainan, the great-grandson of Adam, was entombed on their island. Prior to the Flood, the tradition went, a high tower was situated over the sepulchre, protecting it in a remarkable way. Anyone who approached the tomb was struck dead by a flash of lightning that was discharged from the top of the tower. Of course, all was destroyed by the Flood, but the story of the tower and the tomb had been perpetuated by every generation inhabiting the island since the great catastrophe.

What makes this tradition even more intriguing is that Cainan was a 'craftsman', the inventor of many crafts. According to Bible chronology he died around 3819 BC, which means that he lived for nearly a century following Tubalcain's discovery of the art of metallurgy. In keeping with his craftsmanship, Cainan, in the crowning years of his life, may have combined Tubalcain's knowledge of the properties of metals with his own ingenuity, and become the first man to discover and utilize the power of electricity. This certainly is not an illogical assumption when we have evidence that electricity was used after the Flood by craftsmen, the gold and silversmiths of Babylonia and Persia.

Another item concerning pre-Flood electricity comes to us from a fragment of a Sumerian text cited by the well-known archaeologist S. N. Kramer in his book *History Begins at Sumer* (p. 200). The quoted text speaks of 'Ziusudra, the King, the Preserver of the Seed of Mankind' and how he constructed a 'huge boat which was tossed about' in a flood that overwhelmed the land. Ziusudra is identical to Berosus Xisuthros and the older Sumerian Utnapishtim. The Sumerian text also mentions that in the preparation of Ziusudra's 'huge boat', the hero Utu brought 'his rays [of the sun] into the boat, in order to give it light.' Utu corresponds to Ubarat-utu in the Weld-Blundell Sumerian list, who was the eighth 'king' of the ten pre-Flood rulers – the counterpart of the Biblical Methuselah. Bible chronologists state that Cainan, the discoverer of electricity, died in 3819 BC; and Noah, the utilizer of the electric *tsohar* in the ark, was born in 3998 BC, which means that the two were not contemporaries and that Cainan was not the one who contributed his discovery to the ark. However, Methuselah was 548 years of age when Cainan died, and since he continued to live another 421 years, he certainly was present during the entire period of construction of the ark. Thus, having been a contemporary of both Cainan and Noah, he may have been the individual, as indicated in the Sumerian legend, who relayed the secrets of electric power to Noah.

Electricity, however, does not seem to have been the only energy source of which the antediluvians had knowledge, for there are a number of out-of-place finds and historical records which suggest that they manipulated a wide range of power potentials.

In Genesis 6:14, Noah was commanded to make the ark waterproof in a specific way: '[Thou] shalt pitch it within and without with pitch.' The word for pitch as it is used here is the Hebrew *kopher* – thought to be related to the Assyrian *kupur* – which means bitumen or asphalt. Now, asphalt is a petroleum product, and as we know, natural petroleum was formed by vegetable and animal remains that were subjected to tremendous heat and pressure. The Creationist geologists believe that this occurred when antediluvian life forms were buried by the Flood. Yet, the Genesis account clearly states that Noah was to waterproof the ark with asphalt, which raises the question, Was a petroleum product such as asphalt in existence *before* the Flood? Yes! It must have been, and since it did not happen naturally, we must assume it was produced artificially, which presupposes a highly advanced knowledge of chemistry, particularly in the area of hydrocarbons. If the antediluvians were knowledgeable in hydrocarbon chemistry and production, then the entire range of petroleum products was within their grasp, from waterproof sealants (the 'pitch' of the ark) to plastics and other synthetic materials. Most important, however, they would have been able to produce machine lubricants and engine fuel.

Is it mere chance that the root word of *chemistry* – chemia – is attributed to *khem*, the ancient name for the land of Egypt, or the land of Kham, derived from the Biblical Ham, one of the

54

three sons of Noah?

We do not know which of Noah's sons transmitted the knowledge of electricity to the succeeding generations, but the fact that it survived the Flood is certain, as modern research into the secrets of the ancients furnishes us with ample evidence.

In 1938, Dr Wilhelm König, a German archaeologist employed by the State Museum in Baghdad, Iraq, was aimlessly rummaging through the basement of the museum when he came upon a find that was to drastically alter all concepts of ancient science. It was a storage box containing a number of two-thousand-year-old clay pots which had been excavated at Kujut Rabua, a village south-east of Baghdad.

At first glance the pots were noticeably unusual. Each one was 6 inches high and housed a cylinder of sheet copper 5 inches high and $1\frac{1}{2}$ inches in diameter. The edges of the cylinders seemed to have been soldered with a 60/40 lead-tin alloy, which is comparable to the solder in use today. The bottoms of the mysterious cylinders were capped with copper discs and sealed with bitumen or asphalt. Another insulating layer of bitumen sealed the tops of the pots and was also used to hold in place iron rods suspended into the centre of the copper cylinders. The rods showed unmistakable evidence of having been corroded by an acid solution, long since evaporated.

With a background in mechanics, Dr König immediately recognized that this configuration of copper, iron and acid was not a chance arrangement, but that the clay pots were nothing less than ancient electric cells. Confirmation of his identification came after the Second World War when science historian Willy Ley, working with Willard Gray of the General Electric High Voltage Laboratory in Pittsfield, Massachusetts, constructed a duplicate model of the ancient clay pot cells. They discovered when they added copper sulfate, acetic acid or citric acid – all of which were well known two thousand years ago – the cells produced between $1\frac{1}{2}$ and 2 volts of electricity. Generation of electric current by the same means was not possible in our modern civilization until 1800.

More such electric cells were found. Four similar clay pots containing copper cylinders were unearthed in a magician's hut near Tel Omar (Seleucia), also near Baghdad. Found with these pots were thin iron and copper rods which may have been used to connect the cells into a series – a battery – in order to produce a stronger voltage. Ten other cells were also uncovered at Ktesiphon – again in proximity to the city of Baghdad – by

Professor E. Kuhnel of the Staatliches Museum in Berlin. These were found broken down into their component parts, as though they had been mass-produced and their manufacturer had been interrupted before assembling the pieces into working batteries!

The ancient batteries found in the Baghdad Museum and elsewhere in Iraq all date from the Parthian period of Persian occupation, between 250 BC and AD 650. However, electroplated objects, which presuppose the use of some form of battery, have been discovered in Iraq in Babylonian ruins dating back to 2000 BC. It would appear that the Persians and later craftsmen in Baghdad inherited their batteries from one of the earliest civilizations in the Middle East.

Electroplated objects were also found in Egypt by the famous nineteenth-century French archaeologist Auguste Mariette. Excavating in the area of the Sphinx of Gizeh, Mariette came upon a number of artifacts at a depth of sixty feet. In the *Grand Dictionaire Universal du 19th Siècle*, he described the artifacts as 'pieces of gold jewellery whose thinness and lightness make one believe they had been produced by electroplating, an industrial technique that we have been using for only two or three years.'

Down through the years, diverse sections of the world have yielded many accounts of bizarre and seemingly unexplainable lights, many of which may well have had their source in electric power.

In West Irian – formerly Dutch New Guinea – is a village near Mount Wilhelmina with a layout of artificial illumination that in brightness equals any system we have in our western world. In a United Press dispatch in 1963 Harold Guard quotes visitors to the hamlet as saying that 'they were terrified to see many moons suspended in the air and shining with great brightness.' Other visitors have described these 'moons' as huge stone balls that began to glow with a mysterious bright light as soon as the sun disappeared behind the tangled overgrowth of the jungle. Mounted on tall pillars, they projected a luminous glow over the entire village. This may be the same phenomenon described in 1601 by Barco Cenenera, who wrote about the conquistadors' discovery of the city of Granmoxo near the source of the Paraguay River in the Planalto do Mato Grosso. He wrote, 'On the summit of a 7¾-metre pillar was a great moon which illuminated all the lake, dispelling darkness.'

We know from the historical record that such secret Hebrew societies as the Kabala preserved the knowledge of electricity as late as the medieval period. Eliphas Levi, in his *Histoire de la*

Magie, records the story of a mysterious French rabbi named Jechiele who was an advisor in the thirteenth-century court of Louis IX. Jechiele, his contemporaries report, often astounded the king with his 'dazzling lamp that lighted itself'. The lamp possessed no oil or wick, and Jechiele placed it in front of his house for all to see. What the lamp's secret source of energy was, the rabbi never revealed.

Another device, one with which Jechiele used to protect himself, was a door knocker that literally shocked his enemies. The thirteenth-century chroniclers tell how he 'touched a nail driven into the wall of his study, and a crackling, bluish spark immediately leapt forth. Woe to anyone who touched the iron knocker at that moment: he would bend double, scream as if he had been burned, then he would run away as fast as his legs could carry him.' It would appear that Jechiele pushed a discharge button that sent an electric current into the iron knocker on his door.

The ancients may have had more sources of light than we can imagine, and there are numerous indications that this was so. When the sepulchre of Pallas was opened near Rome in the early 1400s, it was found to be lighted by a lantern which had kept the inside of the tomb illuminated for more than two thousand years. Pausanias, who lived during the second century AD, writes that the temple of Minerva had a light that could burn for at least a year. St Augustine (AD 354–430) claims that in an Egyptian temple dedicated to Isis a lamp burned which neither wind nor water could extinguish.

Until the invention of electric lighting in 1890, we possessed only candles, torches and oil lamps, light sources that smoked and left sooty deposits on ceilings. No trace of smoke, however, was ever found either in the pyramids of Egypt or in the subterranean tombs of the pharaohs in the Valley of the Kings. It has been thought that perhaps the Egyptians used some complicated system of lenses and mirrors to bring sunlight into the burial chambers, but no remains of any such system have ever been found. A number of ancient tombs, in fact, have tunnels and passageways that are too complex for a mirror system to have brought sufficient light into the inner chamber. The only alternative is that the Egyptians had a smokeless light source. Since the Egyptians possessed electricity to electroplate gold jewellery – as Mariette discovered – they may also have utilized it to illuminate their tombs.

How sophisticated were the Egyptians in their understanding

57

and utilization of electricity? In Room 17 in the Egyptian Temple of Dendera, built during the Ptolemaic period and dedicated to the goddess Hathor, a very mysterious picture is engraved on the wall. Egyptologists have been at a loss to explain the meaning of this picture in religious or mythological terms. Several electronics engineers, however, believe it contains information of a very different nature.

First, to the extreme right appears a box on top of which sits an image of the Egyptian god Horus. On his head is his symbol – also the symbol of divine energy – the disc of the sun. This identifies the box as the energy source.

Attached to the box is the representation of a braided cable, which electromagnetics engineer Professor John Harris has identified as a virtually exact copy of engineering illustrations used today for representing a bundle of conducting electrical wires. The cable runs from the box the full length of the floor of the picture and terminates both at the ends and at the bases of two very peculiar objects. Each of these objects rests on a pillar which Professor Harris has identified as a high-voltage insulator. Each object is also pictured as being operated by two Egyptian priests.

The two peculiar objects in the temple picture look very much like TV picture tubes, an impression which may not be far from the truth, for electronics technician N. Zecharius has identified the objects as Crookes tubes, the forerunners of the modern television tube.

In simplified terms, a Crookes tube consists of a vacuum contained in a glass encasement within which a fluorescent ray of electrons can be produced. When the tube is in operation, the ray originates where the cathode electrical wire enters the tube, and from there the ray extends the length of the tube to the opposite end. In the temple picture, the electron beam is represented as an outstretched serpent. The tail of the serpent begins where the cable from the energy box enters the tube, and the serpent's head touches the opposite end. In Egyptian art, the serpent was the symbol of divine energy.

Now, the temple picture shows one tube on the extreme left of the picture to be operating under normal conditions. But in the case of the second tube, situated closer to the energy box on the right, an interesting experiment has been portrayed. Michael R. Freedman, an electrical and electromagnetics engineer, believes that the solar disc on Horus's head is a Van de Graaff generator, an apparatus which collects static electricity.

A baboon is portrayed holding a metal knife between the Van de Graaff solar disc and the second tube. Under actual conditions, the static charge built up on the knife from the generator would cause the electron beam inside the Crookes tube to be diverted from the normal path, because the negatively charged knife and the negatively charged beam would repel each other. In the temple picture, the serpent's head in the second tube is turned away from the end of the tube, as though repulsed by the knife in the baboon's hand.

When one looks at the temple picture as a whole, every aspect represents an important feature of a serious scientific experiment. The one tube with the straight serpent is the control (or the tube operating under normal conditions, for comparison); the other with the repelled serpent is the experimental tube (or the tube upon which new conditions have been imposed). Even the use of a baboon to hold the knife shows that the Egyptians were well aware of the powerful energies they were dealing with and took no chances by participating directly in the experiment themselves.

The Crookes tube was the forerunner not only of television but also of the fluoroscope, an instrument that uses X-rays for diagnosing internal injuries. We have no evidence as yet that the Egyptians possessed the fluoroscope, but we do have indications that the Hindus and Chinese did.

An Indian contemporary of Buddha, a physician named Jivaka, was given the title King of Doctors about 500 BC. Records tell us that he had a 'gem' which he used for diagnosis, and that when a patient was placed before the gem it 'illuminated his body as a lamp lights up all objects in a house, and so revealed the nature of his malady.'

Jivaka's magic gem disappeared in history, but three centuries later there was discovered in the palace of Hien-Yang in Shensi, a 'precious mirror that illuminates the bones of the body'. The mirror was rectangular – four by five feet – and gave off a strange light on both sides. The view of the organs of the body that the mirror gave could not be obstructed by any obstacle, which would be typical of the penetration power of X-rays.

Is it possible that some of these light sources employed energy-conversion methods like electricity, or could it have been something more exotic? Is it possible that the ancients found ways to harness atomic power in order to light small areas? In our day and age we recognize that atomic power will be an important source of energy for the future, but there are indications that atomic power is not new. Not long ago a surprising find was made in West Africa that sheds new light on how far back in history atomic energy was first (?) released.

It was on 25 September 1972, when Dr Francis Perrin, former chairman of the French High Commission for Atomic Energy, presented a report to the French Academy of Sciences concerning the discovery of the remains of a prehistoric nuclear chain reaction. Perrin's first inkling came when workers at the French Uranium Enrichment Centre observed that uranium ore from a new mine at Oklo, forty miles north-west of Franceville in Gabon, West Africa, was markedly depleted of uranium 235. All uranium deposits in the world today contain 0·715 per cent of U 235, but the Oklo mine uranium showed levels as low as 0·621 per cent. The only explanation that could be given for the missing U 235 was that it had been 'burned' in a chain reaction. Evidence in support of this conclusion surfaced when investigators at the French Atomic Centre at Cadarache detected four rare elements – neodymium, samarium, europium and cerium – in forms that are typical of the residue from uranium fission! Dr Perrin concluded his report with the opinion that the Oklo uranium had undergone a nuclear chain reaction which had been spontaneously set off by natural causes. Since the Oklo uranium deposits were geologically estimated to be 1·7 billion years old, Dr Perrin suggested that this is when the reaction took place, for at that time the uranium would have been at its purest.

Following the publication of Dr Perrin's report by the French Academy of Sciences, however, questions concerning his conclusions were raised by many experts. Glenn T. Seaborg, former head of the United States Atomic Energy Commission and Nobel prize winner for his work in the synthesis of heavy elements, pointed out that for uranium to 'burn' in a reaction, conditions must be exactly right. Water is needed as a moderator to slow down the neutrons released as each uranium atom is split, in order to sustain the chain reaction. This water must be extremely pure. Even a few parts per million of any contaminant will 'poison' the reaction, bringing it to a halt. The problem is that no water that pure exists naturally anywhere in the world!

A second objection to Dr Perrin's report involved the uranium itself. Several specialists in reactor engineering remarked that at no time in the geologically estimated history of the Oklo deposits was the uranium ore rich enough in U 235 for a natural reaction to have taken place. Even when the deposits supposedly were first formed, because of the slow rate of radioactive disintegration of U 235, the fissionable material would have constituted only 3 per cent of the deposits – far too low a level for a 'burn'. Yet a reaction *did* take place, suggesting that the original uranium was far richer in U 235 than a natural formation could have been.

So what remains is evidence of a nuclear reaction that cannot be explained by natural means. If nature was not responsible, then the reaction must have been produced artificially. Is it possible that the Oklo uranium is the residue from an antediluvian reactor that was destroyed by the Flood and redeposited in West Africa?

Physicist Frederick Soddy made this significant statement concerning knowledge of atomic physics in ancient myths and legends on page 182 of his *Interpretation of Radium* (New York, 1920): 'One is tempted to enquire how far the unsuspected aptness of some of these beliefs and sayings to the point of view so recently disclosed is the result of mere chance or coincidence, and how far it may be evidence of a wholly unknown and unsuspected ancient civilization of which all other relics have disappeared. It is curious to reflect, for example, upon the remarkable legends of the philosopher's stone, one of the earliest and most universal beliefs, the origin of which, however far back we penetrate into the records of the past, we probably do not trace to its real source. The philosopher's stone was accredited the power not only of transmuting metals, but of acting as the elixir of life. Now, whatever the origin of this apparently meaningless jumble of ideas may have been, it is really a perfect but very slightly allegorical expression of the actual present views we hold today. It does not require much effort of the imagination to see in energy the life of the physical universe and the key to the primary fountains of the physical universe today known to be transmutation. Is, then, this old association of the power of transmutation with,' he concludes, 'the elixir of life merely a coincidence? I prefer to believe it may be an echo from one of many previous epochs in the unrecorded history of the world, of an age of men which trod before the road we are treading today, in a past possibly so remote that even the very atoms of its civilization literally have had time to disintegrate.'

Contrary to what orthodox historians would like to admit, our ancient ancestors seem to have inherited an extremely sophisticated knowledge of metalworking from an earlier civilization. Not long ago, pre-Inca Peruvian ornaments and other objects made out of platinum were discovered. This poses a serious problem, because in order to melt platinum, a temperature of

61

about 1755 degrees Celsius must be reached. We have no satis-factory answer to the question of how the ancient Peruvians were able to produce such a heat.

A few years ago a metal belt-fastener with open-work orna-mentation was found in China, in the burial site of the famous general of the Chin dynasty, Chou Chu, who lived from AD 265 to 316. The fastener was examined by the Institute of Applied Physics of the Chinese Academy of Sciences and by the Dunbai Polytechnic. Their analysis showed that the metal of the fastener was an alloy of 5 per cent manganese, 10 per cent copper – and 85 per cent aluminium.

Aluminium supposedly was not discovered until 1803 and not produced successfully in pure form until 1854. Today, the process of extracting aluminium from bauxite is very complicated and involves the use of a Reverbier oven, refraction chamber and regenerator, as well as electrolysis and temperatures exceeding 950 degrees Celsius.

The question is, Where did the Chinese acquire these elements of twentieth-century technology in the third century? It is possible that they may have even possessed methods of producing aluminium which are still unknown to us today.

The ancient Palestinians seem to have specialized in the perfection of metal-hardening techniques. Professor Clifford Wilson, while working for the Australian Institute of Archae-ology, made this observation concerning a Palestinian bronze statue of Baal. One leg of the statue was missing, and when metalworkers were commissioned to add a modern one, they were surprised to find that they could not duplicate the original bronze. To their dismay and frustration, it was harder than any they could make.

Ancient castings of large pieces as well as evidence of advanced hardening techniques are also found in other parts of the world. In the courtyard of Kutb Minar in Delhi, India, stands the Ashoka Pillar, a column of cast iron weighing approximately six tons and standing twenty-three feet eight inches high, with a diameter of sixteen inches. The column had stood in the temple of Muttra, capped with a Garuda, an image of the bird incarnation of the god Vishnu. But Moslem invaders destroyed the Garuda and tore the column from its original setting, re-erecting it in Delhi in the eleventh century. How long it had been at Muttra we are not certain. It bears the inscription of an epitaph to King Chan-dragupta II, who died AD 413, signifying that it is perhaps fifteen hundred or more years old.

The iron pillar poses a real mystery, not only because of its immense size, presupposing a sizeable casting job, but because of its age. Under the Indian tropical heat and monsoon downpours, a normal piece of iron manufactured in about 413 would have corroded and disappeared long ago. The Ashoka iron column, however, shows only traces of rust, and its existence after one and a half millennia is a testimony to a sophisticated unknown science which the ancients must have possessed.

Another remarkable iron column exists at Kottenforst, a few miles west of Bonn, Germany. Locally it is known as the Iron Man. It has the appearance of a squared metal bar, with four feet ten inches above ground and an estimated nine feet beneath the surface. The iron column was first mentioned in a fourteenth-century document where it was described as marking a village boundary, but there is evidence that the column is much older. Associated with the Iron Man are an ancient stone walkway and the remains of an aqueduct which runs straight towards the column. Like the iron pillar of India, the Iron Man of Kottenforst shows some weathering but very little trace of rust.

While some evidence points to the antediluvians having had a technological level that matched our own, there are also serious indications that in certain areas they entered a sphere of knowledge which has scarcely been nudged by our present-day science.

One of the greatest enigmas in the world involves the Pyramid of Cheops, better known as the Great Pyramid, located on the west bank of the Nile at Gizeh, a short distance from Cairo. It was built in the early Old Kingdom approximately 800 years after the Flood and is purported to be the final resting place for the pharaoh, but no concrete evidence of this has ever been found. The role of the pyramid as a tomb has never totally satisfied most investigators and chroniclers. It has often been stated that this man-made mountain of 2,300,000 stone blocks must have served another purpose than as merely the mausoleum of an ancient ruler. From the fourth-century AD Roman historian Ammianus Marcellinus to the ninth-century Arab savant Ibn Abd Hokem, the writers record the legend that deep within the interior of the pyramid mass were secret rooms containing the lore of a forgotten civilization. The various chambers within the pyramid have since been discovered and fully (?) explored, but nothing of any significance was ever found in them – not even the mummy for which the pyramid was built. We know now, however, that there was a grain of truth in the old legends, for the knowledge was not to be found in any hidden chamber;

rather, it is believed that the pyramid itself is the knowledge.

The mysterious pyramids have been visited by many scientists in the past and present, and a number of these men have noted unusual phenomena associated with the Great Pyramid of Gizeh. Not long after the turn of the century, the British inventor Alexander Siemens travelled to Egypt to see the pyramid and, accompanied by an Arab guide, climbed to the very top. On their reaching the summit, the guide called his attention to the curious fact that whenever he raised his hands with the fingers outstretched, he would hear a ringing in his ears. Siemens followed the Arab's example, but instead of hearing anything, he felt a distinct prickling sensation. Guessing there was something electromagnetic about what was happening, he quickly took a newspaper he had brought with him, moistened it with the contents of a wine bottle, and then wrapped the paper around the empty bottle. In this manner he had quickly manufactured a Leyden jar – a device which accumulates electrical energy. After holding the apparatus over his head, he realized that it became increasingly charged, to the point that sparks began to fly out of it. The Arab guide, who knew nothing about electricity, accused his tourist companion of witchcraft and attempted to seize Siemens by the arm. At that instant, Siemens lowered his spark-shooting bottle towards the man, giving the Arab such a shock that he was thrown to the stones upon which the two stood.

On recovering, the terrified guide scrambled down the treacherous building blocks of the pyramid with scarcely a backward glance, and was never seen again.

Siemens concluded that for some reason the pyramid was discharging a powerful flow of electromagnetic current . . . the 'why' of which he could not answer.

A more recent experiment conducted at the Great Pyramid's sister structure, the neighbouring Pyramid of Chephren, brought the enigma of the energy flow to worldwide attention. In 1968 a group of scientists from the Ein Shams University near Cairo conducted a million-dollar experiment to measure the cosmic rays passing through the pyramid. The goal was to determine whether any undiscovered chambers still existed in the pyramid, for as the cosmic rays strike the pyramid uniformly from all directions, they should, if the pyramid was solid, be recorded uniformly by the detection equipment. If there were vaults, however, the detection equipment would then show a different strength for those areas.

For twenty-four hours a day for more than a year, magnetic tapes faithfully recorded the cosmic rays received by the detectors. Finally, at the termination of the experiment, the tapes were taken to the Ein Shams University to be analyzed by the IBM 1130 computer. The result was absolute chaos! Where there should have been a relatively uniform reading among all the tapes, the computer printouts that plotted the cosmic-ray patterns revealed that the readings were different from day to day.

Dr Amr Gohed, director of the experiment, was quoted in the London *Times* of 14 July 1969, as saying,

'This is scientifically impossible. There is a mystery here which is beyond explanation . . . Call it what you will, occultism, the curse of the pharaoh, sorcery or magic – there is some force that defies the laws of science at work in the pyramid.'

Perhaps the most significant experiments concerning 'pyramid power' were carried out several years ago by a Frenchman named M. Bovis. Arriving at the Great Pyramid in the stifling heat of the day and eager to escape the suffocating temperature, he ventured into the centremost of the pyramid's chambers, called the King's Chamber. Ferreting among the accumulated garbage and debris, he discovered the body of a dead cat. What surprised Bovis was that, despite the humidity of the chamber air, the cat's body had not decayed, but instead had mummified. It was completely dehydrated! A credible explanation for this strange occurrence was not readily available, and Bovis was still perplexed about it when he arrived home in France. There he constructed a scale model of the Great Pyramid, with a base approximately one yard on a side. Recalling that the Great Pyramid is one of the most accurately oriented buildings known to engineering – its base is squared only five seconds, or $\frac{1}{720}$ of a degree, off from magnetic north – he likewise aligned his pyramid model with the North Pole and concluded his experiment by placing several dead animals about one-third of the way up inside the pyramid. Without exception, the bodies of the animals did not putrefy but slowly dried out. Whatever organic matter was placed with the pyramid, the same phenomenon occurred. Brain tissue, when set in a regular box, began to decay within a matter of hours, but while protected by the pyramid's structure, it is reported to have remained unaffected for a period of up to two months. It simply became mummified, with a water loss of approximately 75 per cent.

It was not until the 1950s that Bovis's work with pyramid models came to the attention of Kark Drbal, a Czechoslovakian radio engineer from Prague. He repeated Bovis's experiments and experienced the same results. But he went one step further, and the results of that experiment are still puzzling the experts. Drbal decided to subject a dull razor blade to the mysterious power of the pyramid. To his delight, the dull blade was transformed into a sharp one after fourteen days in the pyramid. Familiar with the scientific method, he duplicated the same experiment several times, with the same result – a force acting within the pyramid restored the blade's edge to its original sharpness.

Visualizing a possible commercial market for his discovery, Drbal tried to have it patented, calling it the 'Cheops Pyramid Razor Blade Sharpener'. The Prague patent office, however, did not share his obvious enthusiasm and flatly refused even to consider its merits until its chief technical advisor constructed a pyramid model for himself and tested it with one of his own blades. To his surprise it worked, and as a result the 'Pyramid Sharpener' was patented in Czechoslovakia in 1959 under Patent No. 91304. A small factory was built to produce little six-inch-high cardboard pyramids, but within a short time it was discovered that any type of construction material produced the same effect. As a result Drbal's pyramids are now being made from styrofoam.

Since the introduction of Drbal's sharpening pyramids in 1960, a considerable amount of research has been conducted, both in the West and behind the Iron Curtain, to unlock the secrets of the pyramid's power. Most of this research has been conducted along scientific lines, but recently, as the mystery of the pyramid force has been probed deeper, science has been superseded by the supernatural. In the United States, Canada, Europe and Australia, pyramids are no longer being used to sharpen razor blades but are rapidly becoming instruments of the occult and are known as 'Feedback-Mystic Pyramids'. The prescribed technique, according to the occultists, is to write a statement or a wish on a piece of paper, insert it in a properly north-south-oriented pyramid, and then pray to the forces within the pyramid to grant the request. Those who have dealt with 'pyramid power' in this manner claim that 'something' is indeed answering their prayers to a degree above and beyond mere chance.

But there's more from Czechoslovakia. Robert Pavlita, another inventor, has taken a step beyond Drbal's pyramid and is now experimenting with all types of shapes and combinations of shapes. He has developed what is now known in psychic circles as the 'psychotronic generator', a 'machine' that supposedly is able to store up and run on energy which Pavlita claims originates from the human mind. When the operator simply concentrates on various points of the generator, the machine can attract non-magnetic material to itself, drive small motors placed in a vacuum, purify polluted water, advance the growth of plants, and heal diseases; and in addition to all of this, it is reputed to be able to perform several occult extrasensory operations. The inventor avows that it can read minds, control thoughts, foretell the future, and communicate with entities that reside on another plane of existence.

What makes these 'psychotronic generators' so intriguing is that Pavlita admits – although somewhat reluctantly – that these machines are not of his invention! Rather, he states that he discovered the principle behind these incredible machines from a number of extremely ancient manuscripts located in the Prague library collection, which incidentally houses hundreds of relic writings still waiting to be deciphered and translated. The manuscripts selected by Pavlita were treatises on black magic – more specifically, magic based on a unified occult-technology developed by an advanced civilization antedating Egypt and Sumeria.

'Pyramid power' and the 'psychotronic generators' have a twofold implication. First, their ancient origin and highly sophisticated technology point to their source as having been *before the Flood.* They show that in the final days prior to the Flood the antedilvuians had advanced in knowledge to a point where they crossed the line separating pure science from pure occultism. In some way they had managed to fuse the super-natural with the natural and in the process had destroyed their civilization.

The second implication is far more ominous. The intensity of the research into the mystery of 'pyramid power' and psycho-tronics is now rapidly approximating the same level that the antediluvians reached before the Flood.

Today, both science and occult knowledge are beginning to approach their theoretical limits – aiming for the ultimate in technical and spiritual manipulation. Is it perhaps possible that we are again edging closer to the danger point?

CHAPTER 3

Following in the Tracks of Ancient Explorers

The final events leading up to the Great Flood are still shrouded in the deepest mystery. There simply are no historical accounts other than the Bible story and the Babylonian Gilgamesh epic that can cast adequate light on one of the most mystifying tragedies of the ancient world. Perhaps because of this, these two venerable traditions are to be treasured more than any other account. Taken side by side, the stories reveal the stark terror that swept over the darkening world as the waters increased.

The Gilgamesh epic says, 'And when the storm came to an end and the terrible waterspouts ceased, I opened the windows and the light smote upon my face; I looked at the sea, tentatively observing, all humanity had turned to mud, and like seaweed the corpses floated.

'I sat down and wept, and the tears fell upon my face.' (Lines 128–37)

'And the ark rested in the seventh month, on the seventeenth day of the month, upon the mountains of Ararat.' (Genesis 8:4)

While the convulsive waters of the Deluge swirled around the ark, inside this stronghold existed another world all its own. Floundering on the angry waves for a period of 150 days before finally coming to rest on the mountains of Ararat, the ship provided shelter and refuge for Noah's family and representatives of the animal kingdom. Seven days after the massive door in the side of the ship had been shut tight, the first torrential rains and the initial shattering earthquakes from the depths of the antediluvian seas marked the end of one world and the beginning of another. For a full thirteen months, the survivors lived amidst chaotic destruction, yet they themselves remained totally preserved from the cataclysm, safely inside the self-supporting vessel as it travelled through a hostile environment.

While the dimensions of the ship are still debatable because of the uncertainty of the length of the cubit, the measure used in the Biblical account, most scholars maintain that the ship's length of 300 cubits, width of 50 cubits and height of 30 cubits should be translated to read 450 by 75 by 45 feet.

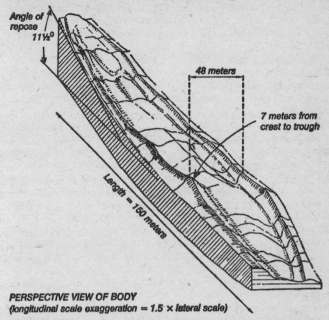

Angle of repose 11½°

48 meters

7 meters from crest to trough

Length = 150 meters

PERSPECTIVE VIEW OF BODY
(longitudinal scale exaggeration = 1.5 × lateral scale)

When studied with the steroplanograph, an aerial photograph revealed some interesting aspects. Dr. Arthur Brandenberger, professor of photogrammetry at Ohio State University, maintains that all of his scientific calculations back the original conclusion that the object was indeed shaped like a ship approximately 450 feet long.

Ancient manuscripts give various interpretations of the dimensions of the survival vessel, the oldest being Origen's description of the ark in *Homilies on Genesis*. He says, 'Judging from the description, I imagine that it had a rectangular bottom and that the walls converged gradually as they rose to the top, where the width was only one cubit . . .'

Origen continues, 'Given the conditions resulting from the rain and the Deluge, a more appropriate shape could not have been given to the ark than this narrow summit which let the rainwater run down, like a roof, and this rectangular bottom flat on the water, keeping the ark from pitching or sinking under the action of the wind and the waves because of the animals' restlessness.'

But what about the specifications? you wonder. Why that specific ratio? Why not, for example, 300 by 200 by 20, or some other ratio?

The specifications mentioned in Genesis speak of a boxlike construction, but not a square box; however, in the Gilgamesh epic, the ark was but a crudely designed cubic vessel having a tendency to turn with each gust of wind as if caught in a gigantic whirlpool.

Noah's ark was different.

Its length-to-breadth ratio of six to one (300 cubits to 50) has great advantage over the cubic contraption of the Babylonian epic. From the standpoint of stability and rolling, the ratio of 6:1 is about as nearly perfect as can be desired. Some of the mammoth tankers of today have a ratio of 7:1. The shipbuilder I. K. Brunel designed the ocean liner known as the *Great Britain* in 1844. The vessel's dimensions were 322 by 51 by 32½ feet – proportions nearly identical to the dimensions of Noah's ark. Whereas Noah's ship ranks as the first of its kind, Brunel had several thousand years of shipbuilding expertise to rely on, but all the accumulated knowledge he could draw from did not give him a better ratio than that of the ark.

Was Noah perhaps acting on advice supplied by a technology that had already reached its summit?

In addition to the speculations concerning the nature and size of its cargo, there are many other aspects of life aboard the ark that deserve close scrutiny. It has often been thought that the only light source on the ark was a window in the roof which permitted the sunlight to penetrate to the interior of the vessel; however, this would not have been very practical. First, any free opening to the outside world would have allowed the water to pour in profusely during the torrential rains of the first forty days; second, with storm clouds brewing above them, there were probably many overcast days and therefore very little sunlight; and third, a series of windows at the top would have lighted only the upper deck, and that dimly, leaving the remaining decks in darkness. If there was a light source, it had to be contained within the vessel itself. Any open fire as a source of light would have been out of the question. The tsohar, powered by electricity (?), may well have been their sole means of illumination. Here at least was something that provided a constant light for the entire Flood voyage and that could have been evenly dispersed on all decks.

But what about air and potable water? Is it possible that the ship contained its own oxygen cycle, using collected plants as the method for air regeneration? Is it possible that Noah actually stored oxygen on the ark to enable him and his cargo to remain alive during the intitial phase of the voyage, when the ship was sealed airtight as a precaution against the raging elements? If the antediluvians were familiar with hydrocarbons such as asphalt, as indicated in the ancient records, then they must have had the chemical proficiency to create and manipulate liquid oxygen. It is therefore not surprising to learn that the manufacture of oxygen was known in the immediate post-Flood era. In the Prince's Library of Ujjain in India, there is a well-preserved document called the *Agastya Samshita* which dates back to the first millennium BC. It contains a detailed description not only of how to construct an electric battery, but also how to utilize this battery to 'split' water into two gases – the electrolysis of water into hydrogen and oxygen. Storing potable water aboard the ark may have generated many problems, for even though the various legendary accounts tell us that Noah lived on the ship for approximately a year, no mention is made of the food and water necessary for the sustenance of his family and the cargo of livestock. It is possible that the ship contained water-tight storage compartments in which these vital supplies were kept, or did Noah perhaps employ a system that pumped water from the sea and filtered out all impurities, thereby making it drinkable?

While speculation reigns when we are attempting to explain the multiplicity of problems Noah must have encountered on his year-long voyage into the unknown, the question of animal cargo as cited in diverse accounts still baffles even the most liberal-minded investigator. Assuming that thousands of different animals did board the ark – Where else would our present animal kingdom originate? – feeding and caring for these animals must have been a job of unbelievable proportions. In addition, a system must have been devised to guarantee that procreation did not occur too often, especially among the more fertile animals, so that the danger of overcrowding could be avoided. It is thought that perhaps the animals' metabolic rates were lowered. If this was achieved, then the feeding and mating would not have been so frequent. Perhaps this was accomplished by artificial means, using the scientific knowledge that Noah and his family possessed. Yet the lowering of their metabolism did not eliminate the need for feeding and care. It is possible that through some form of mechanization, a system of chutes and troughs was used along which food and water were distributed from the storage areas. A similar arrangement may have disposed of animal waste, which was either stored or perhaps ejected from the ark during the voyage. I prefer to think in terms of the latter possibility: knowing when to feed the animals and clean them, etc., demands a knowledge of time; however, during the first forty days of the Flood, when the ship was completely shut up, there was no natural means of keeping track of time. Despite this obstacle, Noah was able to maintain an accurate and detailed diary of events as recorded in Genesis 7 and 8, which indicates that he may have possessed an artificial means of measuring time, probably a mechanical device. This is not unlikely, for at least one example of a timekeeping apparatus from the post-Flood era has been found near Greece. It was discovered in 1900 on the day before Easter Sunday, when sponge divers working off the Greek island of Antikytheros located a sunken Greek cargo vessel filled with bronze statues and other ancient artifacts. From various inscriptions, the shipwreck remains were dated between 80 and 50 BC.

Among the finds brought to the surface was a lump of corroded bronze and wood, which was transported together with the other artifacts to the National Museum in Athens. Several attempts were made to unravel the mystery of the bronze and wood mass, but all were unsuccessful. It was not until 1958 that the unidentified rusting mass came to the attention of Dr Derek J. de Solla Price of Cambridge University. Using an innovative process for restoring oxidized objects, Dr Price was able to salvage bits and pieces of the mass, and by combining these he tried to rebuild the device.

To his amazement, he discovered that the lump had contained parts of an intricate miniature planetarium-computer fashioned from a special bronze alloy. The reconstructed machine was a small box containing more than twenty gearwheels intermeshed in a complex differential system. A crank spindle set the gears in motion at various speeds, turning pointers on three dials that calculated the rising and setting times and phases of the moon, and the positions of the planets Mercury, Venus, Mars, Jupiter and Saturn, all with astonishing accuracy. What's more, it also indicated the time of day.

The Antikytheros clock's construction exceeded the technical capabilities of the Greeks or any other recognized ancient civilizations – yet there it was. Its underlying concept must have originated in an earlier, still more advanced culture, probably that of he pre-Flood world. It is significant that the Antikytheros machine's main function was to calculate the passing of time through simulation of the movement of the heavenly spheres. Could there possibly be a connection with the way in which Noah reckoned his time in Genesis 7 and 8? His calculations also were based on the lunar calendral system, marking off the passing of a solar year.

We do not know, of course, what power source turned the crank spindle of the Antikytheros machine, but it probably was the same source that also lit the tsohar.

We will never know exactly what measures Noah took to preserve life during that year of confinement aboard the ark, but occasionally information will surface concerning the fact that somewhere there exists an account which tends to illuminate this dark period in man's history. In the months prior to the 1950 attempt of the Oriental Archaeological Research Expedition to find the ellusive ark on Mount Ararat, the contents of a curious exchange of letters between Dr Aaron J. Smith, the expedition's leader, and Dr Philip W. Gooch were revealed, and if we ever needed to be shaken into changing the pace of our final arrangements for the trip to Turkey, this correspondence certainly did it. Quoting from what he claimed were ancient records in the possession of a Masonic order to which he belonged, Dr Gooch gave the following information to an unsuspecting Dr Smith: 'There was a living witness on the ground who covered all the fine details of what went on during the Flood and after the Flood until her death in her 547th year,' he wrote to Smith. 'She was God's living witness, Noah's daughter-in-law, the wife of his son Japheth, a student of Methuselah, under whom she was apprenticed, and who taught her all that had preceded the Flood. She was educated in all the history of the human race up to that time. Her book – she called it her diary – is filled with things that occurred from Adam (?) to her death, and seems to me to be the most complete record of early human history ever recorded.

'Many of the problems confronting geologists today can easily be understood after one reads Amoela's diary.

'At her death, dying in the arms of her youngest son, Javan, her diary was placed in her mummified hands in a crystal quartz case, with tempered gold hinges and clasps, and was discovered by a high-ranking Mason in the latter part of the last century. The original and the translation are now in the possession of the Order.'

Subsequent correspondence with Dr Gooch did not result in

concrete evidence concerning the whereabouts of Amoela's diary, and his death shortly thereafter widened the credibility gap even more, as he never revealed the name or chapter of the Masonic lodge.

Was Amoela's diary fact or fiction? It is now evident that we will never know the sources of Dr Gooch's information. One thing, however, we do know. When the great door in the side of the ark was finally unsealed, it opened on to a totally new world, devoid of the life that once had been and ready to receive the life that had been preserved. The rampaging Flood waters had swept away the great technological antediluvian civilization, shattered it into a million pieces and deposited it in the bowels of the earth, away from the sight of the earth's only remaining tenants.

For many centuries Christians have regarded the Mount of Ararat in eastern Turkey as the final resting place of the ark and the spot where the post-Flood civilization began. Only the Genesis account of the tragedy mentions Ararat – all other traditions are silent on this issue. This may be one of the reasons why not all archaeologists agree that the mountain we know today as Ararat is the same one mentioned in Noah's account even though the Turkish name for the mountain means 'Mountain of the Ark', and the Persian name for it is translated 'Mountain of Noah'. The Babylonian legend indicates that the ship was stranded on Mount Nisir, and the Moslem world holds to the view that Mount Djudi is the location. The latter opinion, however, is hardly worth considering, as an increasing number of Islamic scholars have speculated that Mount Djudi may in reality be identical to Mount Ararat.

This controversy, however, does not in the least hinder the Armenians from having their ark traditions and eyewitness accounts.

It was during the aftermath of the ill-fated 1970 Search Foundation expedition that a new development took place, one which could have drastically altered the course of many of the ark-searching activities.

In preparation for the 1970 probe of Mount Ararat, a million-piece mailing programme had been organized for the Search Foundation by Colonel Pak, head of the Korean Freedom and Cultural Foundation, and friend of one of the expedition members. Within weeks after our first contact, 600,000 pieces of promotional mail were dropped in the mailboxes of Americans whose names had been taken from the mailing list of conservative Republicans. Shortly after the first batch went out, a curious phone call was received at the foundation headquarters, and the conversation that followed eventually brought me in touch with an Armenian who may well have been one of the last people ever to see the ark.

It was Mary Board, a semi-retired real-estate broker in Easton, Maryland, who established the first contact.

'There is this old friend of mine in Easton,' she explained hurriedly on the phone, 'an old Armenian, George Hagopian, who says he saw the ark when he was still a little boy. Reading your fund-raising letter brought it all back to me. I've known him for many years, and he has told my husband and me the story of his visit to the ark many times since our first meeting. He says he used to live right at the foot of the mountain. Would you people be interested in talking to him?'

Would we!

In ark research it is always extremely difficult to track down reliable eyewitnesses, for there are too many people with a fertile sense of fantasy who claim to have seen the ark. Up to this point no authentic eyewitness had ever been found. Yet here was a man who was supposed to have lived at the foot of the mountain. To find him living so close by was beyond all expectations.

After some preliminary contacts made by another member of Search, the dean of ark researchers, Eryl Cummings, and I finally met with Mr Hagopian in a smelly diner in Easton, Maryland, and we sat there fascinated, listening to old George Hagopian recall his early days as a poor shepherd boy guiding his restless flocks on the grassy slopes of Mount Ararat.

My tape recorder recalls it all.

'In those days,' George said quietly, his words heavy with an unmistakable Armenian accent, 'my uncle and I used to climb the slope of the mountain to herd the sheep. Everyone who was able used to take their sheep to the green zone of the mountain and graze them there. At daytime there was no problem at all. Just the very idea of accompanying the older men to the slopes of the holy mountain fascinated us – but those nights!'

He closed his eyes for a moment, reliving his boyhood impressions.

'Then the wolves would come, and the bear, and we'd burn the campfires high to keep our sheep safe. The dogs would run and bark all night, checking the flocks and scaring the wolves that would come down the mountain to catch just one little lamb . . .'

George paused, and I quickly threw in a question, hoping to bring the conversation around to the artifact.'

'Were you with your uncle when you first saw the ark?' I asked him.

'Yes, I first went there when I was about ten years old. It must have been around 1902. My grandfather was the minister of the big Armenian Orthodox Church in Van, and he always told me stories about the holy ship and the holy mountain.

'And then one day my uncle said, "George, I'm going to take you up to the holy ark," and he packed his supplies on his donkey, took me with him, and together we started our trek towards Mount Ararat.

'"But, Uncle, that's the holy mountain," I said, pointing to what seemed to be our destination up ahead of us.

'"That's right, George," he said. "Massis is the holy mountain."

'My feet were getting sore, and the donkey kept wanting to go into the wrong direction, but we continued climbing until we got about half-way. Then Uncle took both supplies and me on his back, and we climbed and climbed . . .

'It took us almost eight days from the time we left Van to the moment we got to the place on the holy mountain where both my grandfather and my uncle had said the holy ship had come to rest.

'I guess my uncle took me there that year because it was a year without much snow,' Hagopian said. 'A "smooth year," we called it. There's one of those about every twenty years.

'And then we got to the ark – ' George stopped, groping for the exact words to describe his recollections as clearly as he could.

'I said, "Uncle, it's so dark around here. Nothing but mist. Is this the top of the world? Did the ark rest all the way up here?"

'"Yes," he said, "this is the holy ark. This big ship right in front of you. Let me help you get on it!"

'An immense stone mass loomed threateningly ahead of me. It was like a wall, like a building. It couldn't be; it didn't look like a ship.

'"Is this *really* the ship, Uncle?" I asked, touching the towering object. "This is stone, not wood!"

'"It's the ship all right, Georgie," he answered. "Come and help me. I'll prove it to you."

'He dropped his pack to the ground, and together he and I began to haul stones and huge boulders to the side of the ship. Uncle was a big man, well over six feet tall and very powerful, and within a short time he had stacked a pile of rocks against the side of the ship. Higher and higher we piled them, until he at last told me to stop.

'"George, come here," he said, grabbing me playfully by the arm. "You are going on top of the holy ark." He lifted me up and put me on his shoulders, and together we climbed the pile of rocks. When he had reached the top, his hands grabbed my ankles and he began to push me up.

'"Reach for the top, Georgie!" he yelled. "Grab the edge and pull yourself up."'

A tear welled up in the old Armenian's eyes. He was reliving those nostalgic moments all over again and didn't care whether his emotions were showing. It was his youth he was reliving, and without us as a sounding board it would have been only a voiceless memory.

But now it had life, and he continued:

'I stood up straight and looked all over the ship. It was long, all right. I realize that things always seem larger to a child, but looking back now, I am sure it must have been at least 1000 feet long, and more than 600 feet wide. The height was about 40 feet or more.

'"Look inside the ark," my uncle called up to me. "Look for the holes. Look for the big one. Look inside and tell me what you see."

'I shivered from the cold and from fear and glanced around me. Yes, there was a hole, big and gaping. Was that the one he meant? It looked so mysterious.

'"Uncle, I am scared!" I yelled down to him. "I see a large black hole in the top. Don't make me go in there, please!"

'"Don't be scared, Georgie," he soothed. "There is no one in the ark. It's been empty for a long time. Don't worry about it."

'I peeked into the blackness of the hole, but saw nothing. Then I knelt down and kissed the holy ark.'

'Did you see anything else while you were up there?' I interrupted him. 'Any other distinguishing marks that we might use for identification if and when we locate the object?'

George Hagopian nodded his head affirmatively. His eyes glistened with excitement.

'Oh, yes, many things! There's the moss – that green growth that covers the entire ark. Also when we were there, the top of the ark was covered with a very thin coat of freshly fallen snow, but when I brushed some of it away, I could see the green moss growing right on top. When I pulled a piece off, it looked as if the rock was made of wood. The grain was right there. This

77

green moss, it made the ark feel soft and mouldy. My uncle took his gun and shot into the side of the ark, but the bullet wouldn't penetrate. It just dropped when it hit the side. The whole ark was petrified, turned to rock.'

'Did you see anything on the roof besides one large hole?'

'Yes, I remember *small* holes running all the way from the front to the back,' Hagopian answered. 'I don't know exactly how many, but there must have been at least fifty of them running down the middle, with small intervals in between.

'After describing the holes to my uncle, I asked, "What are these for, Uncle?"

'"They're holes for air, Georgie," he said. "A long time ago there used to be animals and people in the ark, and that's why they had those holes in there. There's one special hole, too, where Noah let the dove fly out."

'"Where did they all go, Uncle?" I asked.

'"They just left, Georgie," he said. "They're gone. The holy ship is completely empty now."

'Uncle pulled his long hunting knife from his belt, and with the heavy handle he chipped a piece from the side of the ark.

'"Uncle, I want to get off," I yelled down to him. "I'm scared. Will you catch me?"

'"Sure," he said, "but don't jump too wildly." So I let myself down the side of the ship until I felt his hands take my ankles in a firm grip. Gently he let me down, and together we went back down the mountainside.

'The first thing I did when I arrived in Van was to see my grandfather.

'"Grandfather, I've been to the holy ark," I said proudly, filled with enthusiasm. "Uncle took me up there. I've been on it. I've looked into the hole!"

'Overcome by the news, Grandfather hugged me, misty-eyed.

'"Georgie, someday you will become a holy man," he whispered, trying to keep his voice from breaking. "You will be a holy man because you've been on God's holy ark."

'He never found out whether his dream for me ever came true, for a few years later he died at Van.'

Intriguing as the story was, the details of the sighting were a bit too hazy, and for another three hours Eryl Cummings and I grilled Hagopian in a friendly way, extracting more facts, especially those that could be useful in our search for the ship's exact location.

By sifting and comparing his various statements, we were able to zero in on precise questions, and George was more than willing to supply the answers from his slowly fading memory.

'I saw the ark a second time,' he recalled, 'I think it was in 1904. We were on the mountain looking for holy flowers, and I went back to the ark, and it looked still the same. Nothing had changed. I didn't get to the top that time, but stayed at the side and really got a good look at it. It was resting on a steep ledge of bluish-green rock about 3000 feet wide. Another thing I noticed was that I didn't see any nails at all. It seemed that the whole ship was made out of one piece of petrified wood. I could even see the grain of the wood, even though the ship had already turned to stone.'

'How about windows and doors?' I asked.

'Oh no! There were no windows in the ship, of that I am certain,' he answered emphatically, his mind's eye roaming the sides of the ship. 'And there was definitely no door in the side of the ship that I could see. No opening of any kind. There may have been one on the side I couldn't see, but that I don't know. That side was practically inaccessible. I could only see my side and part of the bow.'

'What was the shape of the ship? Was it perfectly straight? Was it rectangular, or what?'

George Hagopian paused for a moment before answering, for the waitress had come to pick up our soup bowls. George waited until she had gone.

'The roof was flat, with the exception of that narrow, raised section running all the way from the bow to the stern, with all those holes in it. The sides were slanting outward to the top,' he continued, 'and the front was flat, too. You know, I didn't see any real curves. It was unlike any other boat I have ever seen. It looked more like a flat-bottomed barge.'

'But the *location*, George. Can you describe the location of the ship on the mountain? Are there any specific landmarks?'

The answer came slowly but unhesitantly.

'I do remember that one side of the mountain is impossible to climb,' he explained. 'My uncle and I went through Bayazit, close to the border, and climbed the mountain from the direction of Averbaidzhan. I recall trees and orchards, somewhere between ten and fifteen thousand feet on the mountain. We used to eat the fruit whenever we could. I am sure that the scenery hasn't changed that much. I could take you right to the spot, but at my age, climbing Massis may not be that easy.'

His statement that the wood of the ark had already been

petrified brought us to the subject of wood supposedly discovered on the mountain by an earlier expedition – wood the expedition members claimed had come from Noah's ark.

Hagopian reacted as only he could.

'Listen, son, I don't believe that wood is part of Noah's ark,' he countered, almost interrupting my question. 'The ark I saw was made of wood, *petrified wood*, not wood that can be cut. Also, you said the wood was found at an elevation of approximately 14,000 feet. This proves that it cannot be from the ark, for what I saw was much higher!

'But don't take my word for it. Wait until you have located the real ark, and then you'll see there is no connection at all.'

Five long hours after we first met George Hagopian and interviewed him in the Easton diner, we parted.

'Go find the ark,' he pleaded. 'I can't go, but *you* go! You find the ark. I was there seventy years ago. I know it's there!'

And, kicking the autumn leaves with his shuffling feet, George Hagopian slowly walked back to his frame house.

Since that time practically all of the so-called eyewitness accounts of the ark have been debunked and have been classified as either outright lies or pure fantasy – including that of the young man who claimed to have seen the actual body of Noah hidden somewhere in the vaults of the Smithsonian Institution in Washington, DC – but Hagopian's tale has taken on strength as time goes on. Knowing now for sure that the ancient ship is not to be found at 14,000 feet, I have exposed the Hagopian interviews to the electronic circuitry of the Psychological Stress Evaluator, a lie detector that searches for attempts to deceive by examining the various vibrations found in the human voice. Even though George's voice reveals areas of tension, which can indicate a lie in some cases, this light stress appears to be caused by Hagopian's emotional struggle in attempting to recall his childhood memories. Basically the Hagopian story is sound, and it may well be the first authenticated eyewitness account on record. It certainly fits in with the traditions of the Armenian nation.

Legend and traditions have woven a tight net of folklore about both the slopes of the mountain and the area surrounding it. Nakhichevan, a city south-east of Ararat, was still called Apobaterion – 'place of departure' – during the time of the historian Flavius Josephus. Its present name means 'the place where Noah landed'. Old-timers in that city will proudly direct the way to the traditional gravesite of the revered patriarch. Aghuri, a small town nestled on the slope of the mountain, is the time-honoured place where Noah is thought to have planted his first vineyard after disembarking from the ship. Equally interesting is the fact that Erivan, the capital city of Soviet Armenia, translates as 'the place of the first appearance'.

At various times in history, 'Armenia' and 'Ararat' have been used interchangeably to describe the same area, and even today, Ararat is the name of the province in which the famous mountain is located. Perhaps it is because of the many traditions that have attached themselves to Mount Ararat that the Armenians have always referred to the mountains as the 'Mother of the World'.

Can there be something to this? Are there perchance factors that strengthen this traditional view of Ararat as the spring-board of the post-Flood culture? Concerning the Genesis 10 record of the dispersion of tribes and nations in the dawning days of Middle East history corroborating this 'Mother of the World' concept, Professor W. F. Albright, internationally recognized as one of the leading authorities on Middle East archaeology, says, 'It stands absolutely alone in ancient literature without a remote parallel even among the Greeks . . . "The Table of Nations" remains an astonishingly accurate document . . . [It] shows such remarkable "modern" understanding of the ethnic and linguistic situation in the modern world, in spite of all its complexity, that scholars never fail to be impressed with the author's knowledge of the subject.'[1]

The list he refers to mentions the descendants of Noah, the offspring of his three sons. It gives the first generation of descendants of each son, and, what is more important, it lists the names, which often provide us with clues to their history and dwelling place. The first and second generations left their mark in Egypt, Palestine, Asia Minor, Assyria, Phoenicia, Armenia, the Persian Gulf region and lands in between. The third generation (c. 3230–2780 BC) moved into Europe, Spain, southern Arabia, Lower Egypt, Upper Egypt, the Black Sea region, and Babylonia. The fourth generation (c. 3096–2647) made swift moves into the area presently called Yemen, the land that subsequently was known as the home of the Queen of Sheba. When the fifth generation (3001–2597 BC) arrived on the scene, the record tells of the descendants of Eber, meaning 'pilgrim, migrant', the father of a widely scattered people called Habiru. Very little is known about the individual accomplishments of these people until the fifth generation is reached. Peleg (2867–2528 BC), whose name means 'division, a measurement,' is then mentioned, for

Genesis 10:25 states, '. . . the name of one was Peleg; for in his days was the earth divided . . .'

It is very apparent from the generation list of the sons of Noah that the post-Flood peoples spread rapidly across the surface of the earth. In just the second generation, the grandchildren of the patriarch had settled in lands from Iran to Spain, from northern Europe to Ethiopia. The following generation and their offspring were of course even more widespread. It also becomes obvious that in order for the Genesis 10 genealogy list to have been composed, there must have been an advanced degree of communication among all these people. Someone living during the colonizing of these distant lands had the ability to correspond with all the descendants over a relatively long period of time – otherwise the composition of such a detailed listing as the 'Table of Nations' would have been impossible. This communication between remote regions presupposes an early knowledge of geography. In fact, there is ample evidence that not long after the Deluge, the descendants of Noah carried out an extensive survey of the entire globe, mapping and charting every continent!

The evidence for this post-Flood survey of the earth has been preserved in a number of medieval and Renaissance maps which are extremely accurate – so accurate that the longitude and latitude measurements, as well as the knowledge of the earth's surface that is revealed, extend far beyond the capabilities of the early historical cartographers. These cartographers admit – and there is intrinsic proof of this in the maps – that their maps were copies of still older maps whose origins were lost in antiquity.

One map in particular that has received considerable attention is the Piri Reis chart of 1513. Piri Reis, whose actual name was Ahmet Muhiddin, not only distinguished himself as a captain in the Ottoman fleet of Suleiman the Magnificent, but was also an itinerant map maker and collector. In the most famous of his atlases, the Kitabi Bahriye, and in the notations on his 1513 chart, he revealed that he drew his maps from a composite of twenty older maps. Eight of these maps, he claimed, were from the time of 'Alexander, Lord of the Two Horns', i.e., Alexander the Great. He secured other maps from a captured Spanish sailor in 1501 who told Piri Reis that he had been on Christopher Columbus's three voyages to the New World. In exchange for his freedom, the sailor gave the Turkish captain a number of charts which Columbus had used in locating the islands of the Western Hemisphere. Columbus had, in effect, only rediscovered lands which someone else had charted centuries before.

The Bahriye Atlas remained in use after Piri Reis's death in 1554, but his chart of 1513 was lost until 9 November 1929, when Malil Edhem, director of the Turkish National Museum, was cleaning out debris from the Topkapi Palace in Istanbul and discovered fragments of the old map.

The map received scant publicity when it was found, but copies were sent to various prominent museums. It was not until 1956 that a visiting Turkish naval officer gave a copy of it to the US Navy Hydrographic Office in Washington, DC, where Captain Arlington H. Mallery subjected the map to a comprehensive analysis.

The first extraordinary feature about the map Mallery noted was that it showed South America and Africa in correct relative longitude. In the sixteenth century, when the map was drawn, longitude was found only by guesswork. It was another two hundred years before the correct longitudinal relationship between the two continents was established!

Even more startling, however, was Mallery's discovery that the map accurately showed the coastline of Queen Maud Land in Antarctica – even though the map was drawn in 1513, and the southern continent's existence was not verified until 1819! But there was more. Mallery found that the islands and bays of the depicted coastline are the same as they appear below the antarctic ice sheet, as recently revealed by seismic echo soundings.

In 1957 the map was presented to Reverend Daniel Lineham, S.J., director of the Western Observatory of Boston College, who had participated in an expedition to Antarctica. After careful examination, Lineham reached the same conclusion as Mallery: the Piri Reis map pictured, in great detail, regions scarcely explored today, including an antarctic mountain range that remained undiscovered by modern researchers until 1952. The unavoidable conclusion was that Piri Reis must have possessed charts drawn by someone who had mapped Antarctica *before* the ice cap covered the southern continent. The Piri Reis map could not have been a hoax, for no one in 1929, let alone in 1513, could have reproduced the geographical knowledge this chart contained.

Following a radio broadcast about the map made by cartographer Walters of the US Navy Hydrographic Office in Washington, DC, and Mallery and Lineham, the subject came to the attention of Professor Charles H. Hapgood. Working with mathematician Richard W. Strachan and students at Keene

State College, Professor Hapgood conducted a most meticulous cartographical analysis of the map and other charts of the Renaissance. Professor Hapgood's examination resulted in a number of startling observations, each one of which augments the mystery of the map's origin.

1. The centre of the Piri Reis map is located at the intersection of the meridian of Alexandria – thirty degrees East longitude – and the Tropic of Cancer. Because all the ancient Greek geographers based their maps on the meridian of Alexandria, its use as a centre on the Piri Reis chart confirms Reis's statement that a number of the source maps he used dated back to the Alexandrian period.

2. Another indication of Greek influence in the map was the discovery that the map projection was based on an overestimate of $4\frac{1}{2}$ per cent in the circumference of the earth. Only one geographer in the ancient world had made that overestimation – the Greek Eratosthenes.

When the Piri Reis map grid was redrawn to correct the Eratosthenes error, all existing longitude errors on the map were thereby reduced to almost zero. As Hapgood noted, this could only mean that the Greek cartographers, when they prepared their maps using the circumference of Eratosthenes, had before them source maps which had been drawn without the Eratosthenes error – in fact without error at all! The conclusion is obvious: the geographical knowledge which Piri Reis incorporated into his 1513 map ultimately originated not with the Greeks but with an earlier people who possessed a more advanced science of map making than even the Greeks!

3. The map as a whole reveals a remarkable accuracy of longitude and latitude measurements. In Piri Reis's day, instruments enabling a navigator to find correct longitude were nonexistent. Not until the invention of the chronometer in 1765 were accurate longitude readings possible. Determination of latitude, however, involves precise astronomical observation, but conspicuous differences are evident when it is done by trained men rather than by adventuresome explorers. On his first voyage to the New World, for example, Columbus made no longitudinal measurements and attempted only three for latitude – which incidentally were all incorrect. For almost one hundred years after that famous voyage, European map makers, using the guesswork of the explorers, placed such large islands as Cuba and Hispaniola above rather than below the Tropic of Cancer! In contrast, not only are the Caribbean, Spanish, African and

South American coasts on the Piri Reis map in correct positions relative to each other, but even such isolated land areas as the Cape Verde Islands, the Azores, and the Canary Islands are accurately situated by latitude and longitude – the first two without error and the last within less than a degree. Hapgood commented that there simply is no way to explain the sophistication of the Piri Reis map in terms of the comparative ignorance of sixteenth-century cartographers. The map bears irrefutable testimony to a scientific achievement far surpassing the abilities of the navigators and map makers of the Renaissance, the Middle Ages, the Arab world, or any of the ancient geographers. It is the product of an unknown people antedating recognized history.

4. The Piri Reis chart depicts the Caribbean region at right angles to its normal (Mercator) position, and South America appears stretched out. Hapgood contends that the original source maps from which the Piri Reis map was made must have been drawn using a circular grid based on spherical trigonometry, with the focal point situated in Egypt. Testing this hypothesis, the Hydrographic Office of the US Navy drew a modern map using the same grid, and in such a construction the Caribbean indeed appeared at right angles and South America was elongated. This type of circular projection was not fully developed in Europe until centuries after the map was drawn. Piri Reis revealed his unfamiliarity with such a projection by treating the land area of the original as a flat Mercator-type relief and shifting and splicing the original grid to compensate for the curvature! The Piri Reis map also shows islands and several locations along the Central and South American coast which were either briefly explored but not accurately positioned or not discovered at all prior to 1513. These include the Isle of Pines, Andros Island, San Salvador, Jamaica and others. Further down the coast of South America, the chart shows the mouths of the Amazon and the island of Marajo, correctly shaped and perfectly located in longitude and latitude.

Undoubtedly, the most intriguing feature of the Piri Reis map is the coastline of Antarctica, showing the region of Queen Maud Land. Modern seismic maps disclose that this coast is a rugged one, with numerous mountain chains and individual peaks breaking through the present levels of ice. The Piri Reis map shows the same type of coast, but without the ice. In one instance, Mallery discovered two bays on the Piri Reis map where the seismic map showed land; however, when the experts were asked to check their measurements, they found that the sixteenth-century map was correct after all.

What is the ultimate conclusion of the cartographers? Professor Hapgood and others see no way of reconciling the cartography of 1513 with the data on the controversial Piri Reis map concerning the geography of Antarctica. They concur that the chart indicates that someone possessing measuring techniques which were not employed in Europe until the nineteenth century mapped Antarctica before the continent was covered with ice. Core samples taken in the Ross Sea off the Antarctic coast in 1949 by the Byrd Expedition reveal that there was indeed a time in the distant past when fine-grain sediments were deposited, indicating an ice-free coast and rivers that conveyed silt down to the sea.

Surprisingly enough, the much-analyzed Piri Reis map is not the only map to evince a futuristic knowledge of the earth in remote history. The Orontius Fineus map of 1531 shows rivers in Antarctica where today mile-thick glaciers flow; the Hadji Ahmed map of 1559 depicts the Ice Age land bridge that existed between Siberia and Alaska. The Zeno brothers, in 1380, may have accurately pictured the topography of Greenland below the northern ice-cap, while the Andrea Benincasa map of 1508 indicates that northern Europe was covered by the furthest advance of the Ice Age glaciation.

The only realistic conclusion one can reach on the basis of the accumulative evidence of the medieval maps is that they all have their origin in source maps constructed by an advanced civilization antedating any of the known ancient cultures. Years before the Egyptian, Babylonian, Greek and Roman civilizations existed, at a time when the Antarctic and Arctic were just beginning to feel the advance of the unyielding sheets of glacial ice, this unknown culture possessed a knowledge of cartography comparable to what we have today. These people knew the correct size of the earth; they used spherical trigonometry in their mathematical measurements; and they utilized ultramodern cartographical projections. In addition to their science, these surveyors must also have had at their disposal an advanced form of technology – instruments, and trained specialists to use them, for measuring longitude and latitude. The pre-ancient civilization of the past, Professor Hapgood concludes, must have been organized and directed on a global scale.

In order to place these findings on this ancient universal survey within the historical framework, as endorsed by Professor

Albright, we must carry our assumptions a little further by saying that this survey had to have been made shortly after the Flood (when the land masses were left in their present forms), but before the ice began to accumulate at the poles.

In Genesis 10:25 we meet a descendant of Noah called Peleg who was given his name because 'in his day was the earth divided'. The usual interpretation of this passage is that it refers to the division of nations; however, it could also mean division as in 'allotment, marking off an area, a measurement'. A more accurate translation of this historical passage could therefore read, 'Peleg, in his day was the earth measured, or surveyed.' Even more perplexing is that the record indicates that there have been others equally involved in this cartographic process. Mizraim, a grandson of Noah, comes to mind as one who may have shared in the responsibility of charting the world. His name means 'to delineate, to draw up a plan, to make a representation', especially in association with measuring distances. Mizraim was the founder of ancient Egypt. It is significant to note that at least two of the Renaissance maps showing advanced knowledge, the Piri Reis chart and the Reinal chart, dating back to 1510, were based on a circular projection with the focal point in Egypt.

A third descendant of Noah who presumably also participated in the mapping of the globe was Almodad, whose name, when translated from the Hebrew, means 'measurer'. In the *Chaldean Paraphrase of Jonathan* there is preserved an ancient tradition which tells that he was the 'inventor of geometry', '*qui mensurbat terran finibus*' – 'who measured the earth to its extremities'. Almodad is regarded as a progenitor of the southern Arabians. Is there a connection between him and the fact that many of the Renaissance maps revealed peculiarities of the earth's geography which were first noticed by the Arabs, when taken from ancient sources never fully identified?

The relationship between Peleg, Mizraim and Almodad may be even closer than at first suspected. According to the record, their lifespans overlap, so that the mapping process, covering perhaps the entire period from 2800 BC to 2500 BC, a span of 300 years, was extended over enough years to be total and complete. This conclusion is backed by what we find in the Renaissance maps. It does not leave room for speculation, for among the maps of Antarctica, for example, the Bauche map of 1737 (copied from an older Greek map) shows the continent completely free of ice; the Orontius Fineus map of 1531 indicates that the centre of the continent was beginning to fill with ice when its source maps were drawn, but the Piri Reis chart of 1513 and the Mercator chart of 1569 picture only the Antarctic coast left uncovered by glaciers. It is therefore apparent that Antarctica was surveyed not once but several times, before and during the period the southern polar ice-cap spread over the continent. In the Zeno brothers' map of 1339, Greenland is shown free of glaciers as it was prior to the Ice Age, while Ptolemy's map of the North depicts a glacial sheet advancing across south-central Greenland, and at the same time it shows glaciers retreating from northern Germany and southern Sweden. This could only have come from the findings of surveying parties that tracked the areas before, during, and after the Ice Age.

The world contains a treasure of evidence pointing towards unceasing activity on the part of geographers, surveyors and scientifically oriented explorers during the grey dawn of post-Flood development.

Other Evidence of Post-Flood Geographical Surveys

The scope of the surveying techniques developed by the ancients should not be underestimated. The sacred Hindu books, the *Puranas*, refer to direct communication between India and distant places around the world. The Indians were well acquainted with western Europe, which they called Varaha-Dwipa. England was known to them as Sweta Saila, or 'the Island of the White Cliffs'; and Hiranya, or Ireland, as the Irish legends relate, was visited by the Dravidians, a group of men from India. The Irish say that they stayed for only a brief time and had come as surveyors, not invaders. But the Indian books go far beyond western Europe in their recollection. They describe North America, the Arctic Ocean, South and Central America, and other locations. Detailed research into the background of the ancient Sumerians has also provided us with some fascinating information which tends to connect the twelve zodiacal constellations with the characteristics of the lands found in the directions of those constellations. North-north-west of Sumer, towards Capricorn the Goat, is the Caucasus region known in ancient times for its wild mountain goats, and especially for its domesticated goats, which were exported to Sumer. To the north-west is Aquarius the Water Bearer, and in this direction are Asia Minor and the source of the Tigris and Euphrates rivers. In mythology, the special god of a river was always pictured as replenishing the source by pouring water into it. Pisces the Fish was found west-north-west, towards the Canaanite and Phoenician coasts, famous for their fishermen and bountiful catches – and so it continues with the other constellations. There was always a connection between the sign of the zodiac and the lands found in its direction, and when contemplating the geographical knowledge demonstrated by the zodiac and the countries it represents, one must conclude that during a very early period in their history the Sumerians were familiar with lands as distant as North Africa, India, Ethiopia, the plains of southern Russia, even including all of the eastern Mediterranean and western Asia . . . certainly far beyond the accomplishments of a people often regarded as 'primitive' ancients.

Evidence of a World Survey in Egypt

Serious consideration must be given to the involvement of Mizraim with the world survey that was conducted after the Flood. We know from Egyptian history that Mizraim is regarded as the forefather of all Egyptians, and it is significant that the secular records of Egypt testify that from a very early period the Egyptians were indeed knowledgeable about land measurements and practiced sophisticated surveying techniques.

Livio Catullo Stecchini, one of the world's foremost authorities on ancient measures, discovered a peculiar hieroglyph that appeared on all the thrones of the pharaohs, beginning at the Fourth Dynasty. The hieroglyph is composed of knotted ropes symbolizing the union of Upper and Lower Egypt at the thirtieth parallel, where the southernmost tip of the Nile Delta crosses the meridian 31° 30′ east of Greenwich, which appears to have been established as the prime meridian of Egypt in unknown antiquity. At the bottom of the hieroglyph are three pairs of horizontal lines of different lengths, depicting the three sets of values which the Egyptians gave the Tropic of Cancer. The middle line represented the conventional tropical latitude of 24°, the lower line symbolized the actual latitude of 23° 51′, and the top line lay at the latitude of 24° 6′. This last latitude, being 15′ north of the true line, is important because 15′ is half the diameter of the sun, which shows that the Egyptians understood that it is not the centre of the sun but rather its outer rim which must be observed for geodetic survey. Precisely where 24° 6′ crosses the Nile, on the island of Elephantine opposite Aswan, the

Egyptians had an important astronomical observatory.

Many significant cities of Egypt, it appears, were built in relation to the established prime meridian of Egypt and the Tropic of Cancer. The predynastic capital of Lower Egypt, Buto, was located precisely on the prime meridian, 31° 30', near the mouth of the Nile. Memphis, the first capital of unified Egypt, was also placed on the prime meridian, at 29° 51' – exactly 6° north of the Tropic of Cancer. In the Twelfth Dynasty the capital was moved once again, this time to Thebes. A new central meridian for Egypt had been established at 32° 38' east which paralleled the eastern edge of the Nile Delta. Thebes was located where the meridian touched an eastward bend in the Nile at 25° 42' 5" north. What is amazing is that this parallel is almost exactly $\frac{2}{7}$ the distance between the equator and the North Pole.

The survey work of the Egyptians left its mark not only on the land of the Nile but throughout the rest of the ancient world as well. Stecchini has found that such early capitals as Nimrud in Mesopotamia, Sardis in Asia Minor, Susa in Persia, and even Anyang in China were established in relationship to the earliest prime meridian of Egypt. In terms of latitude, Delphi and Dodona, the two most important oracular shrines in early Greece, were also founded in relation to Egyptian measurement, being 7° and 8°, respectively, north of Buto.

Stecchini believes that when Alexander the Great destroyed Heliopolis, the centre of Egyptian science, and replaced it with his own centre at Alexandria, he may have destroyed the last vestiges of Egyptian survey knowledge. Far from being the great men of science they were for so long thought to have been, the later Alexandrian Greek geographers did nothing more than revive in part the advanced science of geography that had preceded them.

Evidence of a World Survey in China

Among the early Chinese we find evidence that they too possessed advanced knowledge obtained from the geographical survey of the world taken soon after the Flood. One of the oldest Chinese literary works that has survived is called the *Shan Hai King*, *The Classic of Mountains and Seas*, a treatise on geography. Its authorship is ascribed to 'the great Yu', who became Emperor in 2208 BC, and the date for the writing of the treatise is approximately 2250 BC – about a century after the death of Almodad, the seventh-generation descendant of Noah who 'measured the earth to its extremities'. For several hundred years after its writing, the *Shan Hai King* was regarded as a scientific work, but during the third century BC, when many Chinese records were re-evaluated and condensed, it was discovered that the geographical knowledge it contained did not correspond to any lands known at that time. Thus, the *Shan Hai King* was re-classified as myth and was relegated to an unimportant position in Chinese literature.

Within the past few years, however, several portions of the *Shan Hai King* have been re-examined, and the information they contain has altered many previous assumptions concerning the treatise. In the Fourth Book, entitled *The Classic of Eastern Mountains*, are four sections describing mountains located 'beyond the Eastern Sea' – on the other side of the Pacific Ocean. Each section begins by depicting the geographical features of a certain mountain – its height, shape, mineral deposits, surrounding rivers and types of plants and vegetation – then gives the direction and distance to the next mountain, and so on, until the narrative ends. By following these clues and the directions and distances provided, much as one would a road map, investigators have discovered that these sections describe in detail the topography of western and central North America.

The first section begins on the Sweetwater River and proceeds south-east to Medicine Bow Peak in Wyoming; then to Longs Peak, Grays Peak, Mount Princeton, and Blanca Peak in Colorado; to North Truchas Peak, Manzano Peak, and Sierra Blanca in New Mexico; then to Guadalupe Peak, Baldy Peak, and finally Chinati Peak, near the Rio Grande in Texas.

The second section describes an expedition over an even more expansive area. It begins in Manitoba, at Hart Mountain near Lake Winnepeg, and proceeds to Moose Mountain in Saskatchewan; it goes from there to Sioux Pass (between Andes and Fairview) in Montana; to Wolf Mountain and Medicine Bow Peak in Wyoming; to Longs Peak, Mount Harvard, and Summit Peak in Colorado; then to Chicoma Peak, Baldy Peak, Cooks Peak, and Animas Peak in New Mexico; then on into Mexico, describing the Madero, Pamachic, Culiacan and Triangulo heights, reaching the Pacific Coast near Mazatlan.

The third section is a tour of the mountains along the Pacific Coast: Mount Fairweather and Mount Burkett in Alaska; Prince Rupert and Mount Waddington in British Columbia; Mount Olympus in Washington; Mount Hood in Oregon; and

Mount Shasta, Los Gatos, and Santa Barbara in California.

The fourth and last section covers several peaks in a small area: Mount Rainier in Washington; Mount Hood, Bachelor Mountain, Gearhart Mountain, Mahogany Peak, and Crane Mountain in Oregon; and Trident Peak and Capitol Peak in Nevada.

Not only is *The Classic of Eastern Mountains* a geographical survey, but the accounts in each section give the observations and experiences of the surveyors, from picking up black opals and gold nuggets in Nevada to watching the seals sporting on the rocks in San Francisco Bay. They were even amused by a strange animal who avoided its enemies by pretending to be dead: the native American opossum.

Other portions of the *Shan Hai King*, specifically the Ninth and Fourteenth books, also describe regions in North America. One notable description given in the Fourteenth Book is of a 'luminous' or 'great canyon', 'a stream flowing in a bottomless ravine', in the 'place where the sun in born'. Anyone who has witnessed a sunrise in the Grand Canyon will know what the early surveyors had seen. Still other parts of the *Shan Hai King*, currently under investigation, are said to be accounts of explorations further to the east, in the Great Lakes and the Mississippi Valley areas.

It is very evident from the accuracy of the geographical details and the personal observations in the *Shan Hai King* that an extensive scientific survey of the North American continent was made by the Chinese almost 4500 years ago.

A World Survey – A World Language

After the initial enquiries into the *Shan Kai Hing* were begun, it was noticed that along many of the routes which the surveyors of North America took, there existed several examples of rock drawings. The most notable are Writing Rock near Grenora, North Dakota, and Writing-on-Stone in Alberta, Canada. Yet another rock script occurs in British Columbia, and petroglyph expert Philip Thornburg was the first to recognize among the stone pictures a carving of a *sisutl* – the Chinese dragon. Thornburg observed, 'There does seem to be an Oriental background to them. Since they are carved in sandstone, it's virtually impossible to say what age they are. I've found some that were buried under a foot of topsoil. Now this wasn't the kind of topsoil that would have washed over them. This was formed there, placing the age of the carving around five to seven thousand years – which is really ancient for this country.' Thornburg discovered one petroglyph on Vancouver Island that had had a hole worn through it by dripping water, proof that it had been there for some time.

William and Mae Marie Coxon, amateur archaeologists, have spent the last decade studying the Canadian and other petroglyphs found around the world. The conclusion of their research has been that at a very remote priod in human history a group of people they call the Stone-Writers left their traces on every continent. By careful comparison, the Coxons discovered 241 special sequences of particular geometric signs and symbols. The distribution of examples of these sequences was 201 in the Middle East, 171 in the Far East, and 131 in the Americas. By dating the petroglyph remains in the Nile Valley to compare with the later Egyptian civilization, the Coxons were able to date the Stone-Writers' appearance as being about 1500 years before the rise of Egypt.

From the drawings themselves, the two researchers were able to describe the Stone-Writers as average to above average in height, wearing short kilts that came to the knees, much like the ancient Egyptian labourers. They must have possessed great strength and endurance to have penetrated into the inhospitable terrain where many of their glyphs were found. The Coxons are convinced that the Stone-Writers were not barbaric hunters or nomads but an intelligent people who were systematic in what they did; the symbols had meaning and purpose in their repetition and locations. The Coxons note, 'They travelled the oceans, or at least the coastlines, and they penetrated far up into the continents along the rivers . . . Along the streams, lakes and ocean shores, they left guide signs to mark the way for others who followed them . . .' The Stone-Writers were thus explorers and geographers, probably the very same explorers and geographers who charted the world after the Flood.

The Coxons' work in symbols is being verified by a number of other researchers. English archaeologist S. F. Hood, after studying tablets at the prehistoric site of Tartaria in Rumania, discovered correlations between the tablet symbols there and those found in Crete, Iraq, Egypt, and the Balkan countries. His conclusion was that a system of signs was used over an extensive area 6000 years ago. N. Vlassa, of the Museum of Cluj, supports these findings with discoveries of his own. Almost identical symbols from the same time period appear at Vinca and Tordos in Rumania, at Troy, and on the Aegean island of Melos. On the basis of his own research and that of his colleagues, Hood believes that a single system of glyphs originated from Iraq or some other country in the Middle East and were disseminated from there over a wide area in a very short time. Oswald O. Tobisch, in his work *Kult Symbol Schrift*, has carried the research a step further and, like the Coxons, sees striking parallels in symbols in Africa, Europe, Asia and America.

Symbols left on rocks and tablets for others presupposes communication by language.

Significant research into language usage in the past has been done by the Irish etymologist John Philip Cohane. More specifically, Cohane has concentrated his efforts over the last several decades on a detailed study of the origins of words in almost every language around the world. He has discovered that many words contain similar root words or root combinations beyond what mere chance would permit. These reappearing roots, Cohane discovered early in his study, all have their origin in the Middle East and either figure prominently in Semitic legend or are found in the Old Testament, notably in Genesis. Cohane commented concerning these widely dispersed root words, 'This is not to say there may not be a more logical, even earlier, point of origin than the Semites, but if so I have not been able to find it. On the basis of the evidence, it would seem that a high percentage of the people of the earth today are far more closely related than is generally assumed and they that are bound together by at least one early blood stream that is Semitic in origin.'[2]

Later in his research Cohane was able to establish that there had been two major dispersals of peoples from the Middle East in the distant past, each group taking with it a previously established group of root words. The second of these was heavily concentrated in a limited area of the world: the Mediterranean basin, Europe, Africa, parts of Asia, the West Indies and Brazil. The first dispersal, however, though its traces are less apparent today than those of the second, covered the entire world, in a very short time, in Cohane's opinion. He says, 'If one puts a charted overlay containing only the first group of names on top of a map of the world and then puts on top of that another charted overlay containing only the second group of names, the most logical conclusion is that, in prehistoric times, instead of one there were two dispersions from the Mediterranean, the first truly worldwide, the second petering out along the eastern coast of the Americas in one direction, [and] in Japan, the

Philippines, Australia, and New Zealand in the other direction. Again, there may be a more logical conclusion to be drawn from the data, but if so, I have not been able to think of one. And again . . . *all the key names in both groups have prominent origin points in Semitic legends and mythology, as well as known Semitic place names.*'[3]

These findings tell us several things. The discovery of symbol and word dispersals from a common point of origin somewhere in the Middle East fully corroborates the historical Genesis record and its story of the dispersal of nations from one point. The worldwide language wave is also indicative of the ancient geographical survey of the world, as established by the maps found in Renaissance times. A great number of the significant roots were found by Cohane in geographical names. For those who seek in this a confirmation of the Biblical story, the Genesis 11:1 description of the conditions following the Deluge is interesting: 'And the whole earth was one language, and one speech.' Cohane's second language dispersal might well be identical with the language disturbance that followed the confusion at the Tower of Babel, as mentioned in Genesis 11:7.

The Reason for a World Survey – The Earth's Magic Lines

It is quite evident that soon after the Flood and before and after the language disorder, between the second and seventh post-Deluge generations (c. 3100–2500 BC), the descendants of Noah undertook a geographic survey and exploration of the entire world's surface, leaving their traces in the form of maps, symbols and place names. This was most assuredly accomplished with knowledge preserved from the antediluvian era, yet why was it done? Why did the ancients undertake such a momentous task? Why did they decide on this type of adventure, while the memory of the global devastation was still fresh in their minds?

There are some obvious explanations. As Noah and his family stepped down from their survival vessel, they looked upon a world totally alien to them. All the familiar landmarks had disappeared. Forests were gone, rugged mountain peaks faced them on all sides, and from the murky waters below rose the foul stench of decay. The earth that they once had known was now wiped completely clean of any previous civilization. It was as if they had landed on another planet.

As the new generations were born and grew up on the foothills of Mount Ararat, their innate curiosity concerning this new land

95

forced them to venture out into the hinterland, to explore for fertile valleys, plains and forests. The record shows that those early generations were well aware that they were the fathers of future nations, for their names often denoted their professions or the geographical areas they occupied. They moved out with stubborn insistence, initiating the first 'land grab' in history, staking out their territories, and when they found an area which was suitable to become the home of a nation, they settled and laid claim to the newly acquired territory, reserving it for their children and their children's children. The valuable resources of the earth had been washed away and laid down in new deposits by the turbulent Flood waters. The natural inclination of the post-Deluge generations would be to search for these treasure-troves. Professor Hapgood suggests still another reason. He believes that the mapping of a continent on such a vast scale, as with Antarctica, requiring much organization, numerous exploring expeditions and many stages of data compilation, must have been motivated by a powerful reason. He feels that economic gain may have been this reason; yet, the exploratory expeditions did more than merely discover and cultivate new areas; they actually divided the earth into parcels of land, with each one bounded by what are now called ley lines.

Until a warm summer afternoon in the early 1920s, there was no indication other than the historical Genesis record that this ever occurred. Alfred Watkins, a merchant whose hobby was prehistory, was riding horseback through the Bredwardine hills near Hereford, England. On reaching the summit of a grassy hillock, he rested, letting his eyes gaze over the tranquil English landscape. Suddenly he saw something he'd never noticed before. Several church steeples were aligned straight across the country-side. Knowing that these churches had been constructed on the sites of prehistoric sanctuaries, he wondered whether it was possible that they had once been linked by an invisible web of lines. While still pondering this question, he suddenly realized that not only ancient temples, but also mounds, old standing stones, crosses, crossroads, sacred trees, moats and sacred wells also stood on the same lines!

Racing home, Watkins painstakingly marked all the ancient sites and monuments he knew from his studies on to a one-inch ordnance map, and even though finding five or six points in alignment would have been beyond mere chance, he found himself confronted with eight, nine and even more points, all stretching out in precisely straight lines! Carrying his initial research a step

Dr. Clifford L. Burdick, Arizona geologist, with a set of petrified tracks left in the Paluxy River bed in Texas by a dinosaur and a human being of formidable dimensions. The human footprints are 15 inches in length, but even though their length is extraordinary, much more significant is the fact that they were found *next* to the dinosaur tracks. It indicates that man and the dinosaur were contemporaneous, not millions of years apart, cutting the evolutionary time-table to shreds.

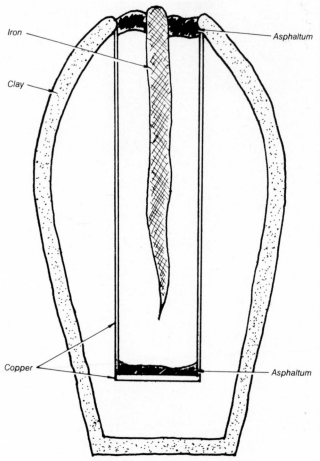

The 2,000-year-old Baghdad battery as sketched by Willard Gray of the General Electric High Voltage Laboratory. Gray's replica, based on the specifications of the unearthed battery, produced between one and a half and two volts of electricity, showing that electric power was in use in ancient times.

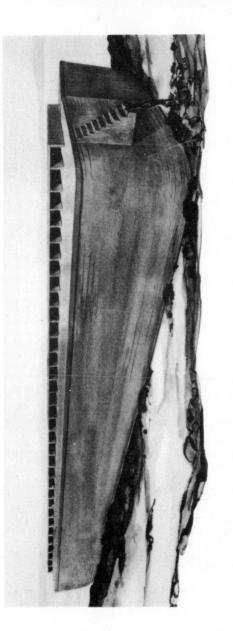

Based on a series of interviews with George Hagopian, illustrator Elfred Lee drew this conception of Noah's ark. This is what Hagopian claims he saw on Mount Ararat when he climbed it as a young shepherd boy.

The strange formations amidst the jagged rocks that make up the mountains of Ararat have given rise to much speculation. Many researchers believe that the little white "object" in the middle of this photograph is the legendary ark of Noah.

Turkish soldiers who took part in the 1960 expedition on their way back to the barracks. The ark is to the right and in back of the soldiers. We are looking at it from stern to bow.

One of the most exciting expeditions in the search for the legendary ark was launched in 1960 as a result of this aerial photograph taken by the Turkish Air Force. On-the-spot investigation confirmed the initial findings that the object had the right dimensions — but there's where all comparisons ended. Some investigators, however, still believe that the shiplike formation is the actual outline of Noah's ark, buried beneath.

George Hagopian, the Armenian who claims to have seen Noah's ark on Ararat. His story bears an amazing resemblance to other accounts, but he places the ship at a much higher elevation than do many ark-researchers.

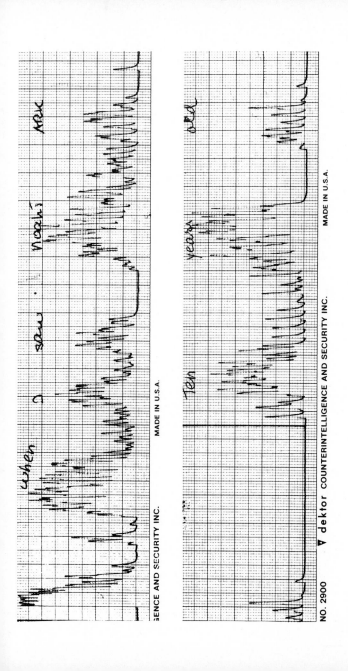

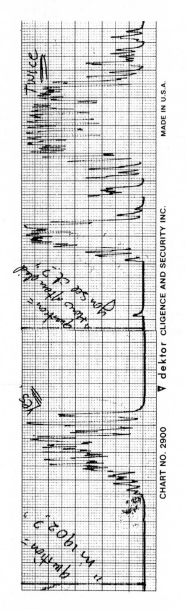

The most interesting and most reliable of all eyewitness accounts of a discovery of the remains of Noah's ark was given to members of an ark-searching team in 1970 by George Hagopian. George, who was born in Armenia, claims to have climbed the mountain when he was a ten-year-old boy and to have seen and climbed onto the ark in the company of his uncle. Careful examination of his tape-recorded story on the Psychological Stress Evaluator (PSE) fully backs his account. This and the following graphs are indicative of a truthful account.

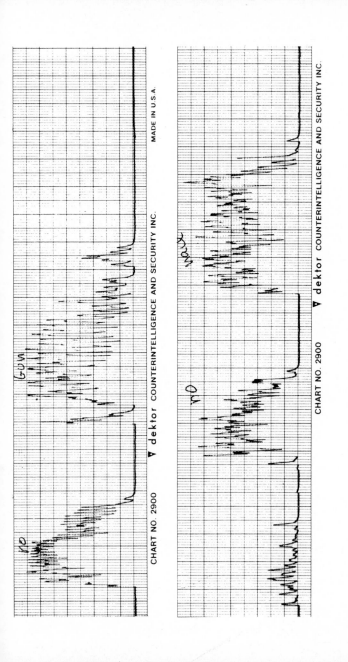

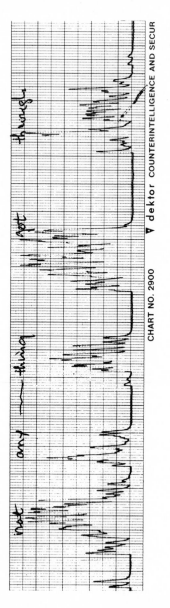

Explaining the stonelike quality of the remains of the ship, Hagopian stated that "no gun, no nail, not anything got through," referring to his uncle's attempt to shoot a bullet into the ancient hull. Here, too, Hagopian is totally stressless.

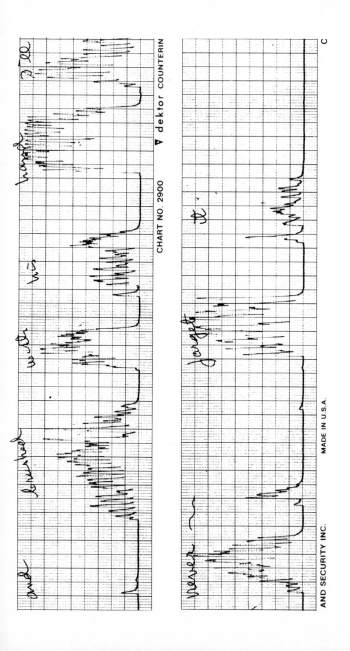

CHART NO. 2900   ▼ dektor COUNTERIN

AND SECURITY INC.   MADE IN U.S.A.   C

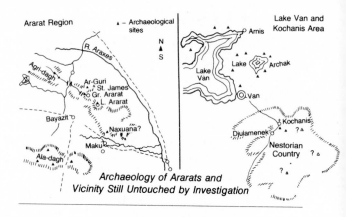

Archaeology of Ararats and
Vicinity Still Untouched by Investigation

Archaeological Sites and Data on Mt. Ararat
and the Surrounding Region
by Col. Alexander A. Koor

Part One: Archaeological Sites

There are inscriptions located as follows (sufficient investigation would reveal many more than these): **1.** *On the Araratsky Pass (just northwest of Greater Ararat), high on the eastern wall, there are several inscriptions.* **2.** *On the summit of Lesser Ararat there are several tombs and some inscriptions on the stones (2' × 3'), not yet deciphered.* **3.** *On the Ala-dagh [mountain] southwest from Bayazit.* **4.** *At Lake Archak.* **5.** *North and northeast of Lake Van.* **6.** *In the Nestorian country (home of ancient Assyrians).* **7.** *Close to Maku city, on the southern slope of the Lesser Ararat, the ruins of the ancient town Naxuana; here is the tomb of the Patriarch Noah, according to Armenian historian Vartan.* **8.** *On the northeastern slope of Greater Ararat, the ruins of the ancient site of Ar-Guri.* **9.** *Not far from Ar-Guri, the ruin of the St. James Monastery, destroyed by an earthquake in 1840. (Many ancient books and relics of Noah's epoch were lost here.)* **10.** *On Agri-dagh [mountain northwest of Greater Ararat], a very old road (called Colchis Road). This area should be searched.*

The inscriptions are in the following languages: Hettit, Assyro-Babylonian (cuneiform), and Arabic-Tartar. There are some Sumerian cuneiform inscriptions around Lake Van.

Perhaps fifty or more other archaeological sites in the Ararat region have never been investigated—the ruins of ancient villages, towns, buildings, tombs, caves, etc.

Russian colonel Alexander A. Koor's sketch as it was submitted to the Sacred History Research Expedition in 1946. It shows the archaeological sites still to be explored in the region of Mount Ararat. The ark-searching expeditions have thus far neglected these sites.

Sacred History Research Expedition

FIRST MESSAGE FROM THE ARARAT REGION
By Col. Alexander A. Koor

Some inscriptions of Karada (Kara-dagh?)

Some inscriptions which I found and copied in the year 1915 in the region of Mt. Ararat:

1...............................?
2...............................?

I translated the pictorial signs and they give me an accurate explanation of the origin of the words MAGOG, GOG, etc. (Ezekiel 38:2 and 39:1) This explanation also upholds the truth and accuracy of Chapter 10 and of Genesis. In short, the words Magog, Gog are words of the Sumerian tongue, which shows the ancient origin of the words. Jews incorporated the words through the Assiro-Babylonians. The word MAGOG consists of two words: MA and GOG.

The word MA is a pictorial sign which is and in cuneiform:

It means: (1) "origin, (2) "the seeds of the waters (FLOOD)".

The word GOG, his pictorial sign is and in cuneiform:

It means: (1) "the Word of GOD" or "by the Word of God."

(2) "thrown" or "banished by God."

Having two meanings:

-"originating from (by) the Word of God" and "the seeds of the waters (FLOOD), thrown by GOD," are both amazingly coincidental with BIBLE texts. Noah came in Mt. Ararat's region with the Flood, that is, with the waters. Likewise, the first meaning: "Originating with the Word of God", is upheld in Chapter 1, Verses 1, 2, 3, The Gospel of St. John.

The word GOG is met with in the text "Ishtar", from the library of the temple Ishtar in Nipur, and also in the text of "7 tablets of the Creation of the World", part 6, p. 11. For linguistical disbelievers of Ancient Ball "Sumerians", pictorial signs, as proof.

Pictorial sign is read : TO KOOR and means: to come to rest on mountains" or "peak". Cuneiform KOOR" means, mountain. (Koora or Kura.)

Pictorial sign is read: TOOB and means: "to filled earth." (TUB-AL).

Cuneiform:

Pictorial sign is read E-GA and means : "the flood, night descended from upon."

The signs:

they may be translated as follows: "These thrown by God, by waters came to rest on the mountains when the night and the waters (flood) descended from upon the earth......."

Since these four pictorial signs alone, through the centuries, were well preserved from two or more lines of damaged inscriptions (possibly by earthquakes), I copied only the four signs. The place where these signs are inscribed is called "Karada". (m. b. Kara-dagh).

Another report submitted by Colonel Koor to the Sacred History Research Expedition in 1946. He was on army duty in the Ararat region around the beginning of World War II and copied many of the inscriptions found in that area.

The Tower of Babel, symbol of the Babel world centre a
mentioned in the book of Genesis. It is supposedly from her
that most post-Flood races originated. Most languages ca
trace their roots to this part of the world.

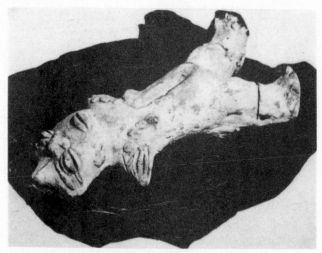

One of the pre-Inca statues showing a Caesarean section, indicating the advanced state of medicine in ancient days.

further, he compared his points to positions on other maps he had marked, and discovered that the lines could be extended for miles and miles, usually ending at a mountain peak or a high cliff. Aided by a friend, Watkins undertook a detailed survey of all England and Scotland and everywhere they found further traces of a prehistoric network of dead-straight alignments that had once extended over the entire island.

Building on the accomplishments of Watkins, Major H. Taylor of the British Army set out, accompanied by a professional surveyor, to do an even more detailed study of the strange alignments. Taylor discovered more landmarks previously not known, or at least not recorded in modern times. His findings were eventually published in a small book entitled *The Geometric Arrangement of Ancient Sites*. But if he thought he'd have a publishing first, he was mistaken, for a year prior to the emergence of his book, a German geographer, Dr Heinish, had already presented a paper dealing with the same discoveries before an international congress held at Amsterdam. Delivering his paper, 'Principles of Prehistoric Cult Geography', he proposed to a hushed audience that at one time in the distant past a magical principle had existed by which holy sites were situated. They were placed, he submitted, on lines that were constructed in relation to the positions of the sun, moon and planets. In addition he claimed he had uncovered evidence that the units of measurement used to construct these lines were, like those of the early Egyptian geodetic surveys, based on simple fractions of the earth's dimensions. He had found examples of these lines not only in Britain but all over Europe and the Middle East. Greatly impressed by the vast extent and accuracy of the construction of these lines, Heinish concluded that they bore testimony to the past existence of a widespread civilization that possessed advanced knowledge of both technology and magic.

Beyond the boundaries of Britain, the lines have been found in nearly every corner of the globe, and strangely enough, associated with them were numerous stories of a flow of magic energy! Ireland has many legends of fairy paths, along which fairies and other spiritual beings are supposed to travel at specific times of the year. Yet today these same 'magic' routes have become well-worn roads and footpaths. In his book *The Fairy Faith of Celtic Countries*, J. D. Evans-Wentz recalls how an old Irish seer explained to him that mysterious currents flow along the paths, but that their exact nature has been forgotten.

Similar research conducted by Xavier Guichard strongly

supported the findings of the British and German investigators. Referring to several old cities in his native France, Guichard said, 'These cities were established in very ancient times according to immutable astronomical lines, determined first in the sky, then transferred to the earth at regular intervals, each equal to a 360th part of the globe.'

Evidence that these lines existed in remote history can be found in ancient literature. For example, in their conquest of the Etruscans, the early Romans noted standing stones set in linear patterns over the entire countryside of Tuscany. Later, during the Latin invasions of Greece, they recounted the fact that 'stone pillars' were found running straight and true along the roads through the hilly Hellenic landscape. The Romans were not particularly surprised at finding these straight tracks, for they had discovered them in almost every country they subjugated: across Europe, North Africa, Crete, and as far west as the regions of ancient Babylon and Nineveh. We now recognize that the Romans' reputation as builders of straight thoroughfares was partly attributable to their simply utilizing sacred lines that existed long before their conquests, and which they then transformed into military and commercial routes. Even today, the Bedouins of North Africa use the line system marked out by standing stones and cairns to help them traverse the desert wastes. When these were stones erected?

The nomads shake their heads when asked this question, for even though they need the markers for survival, they know nothing of their origin.

While the lines have been forgotten in most countries, in other parts of the world the prehistoric line system is still being used. One such system exists in the interior of Australia among the aborigines, who tell of a past age, which they call 'dream time', when the 'creative gods' traversed the country and 'reshaped' the land to conform to important paths called *turingas*. At certain times of the year, they say, the turingas become revitalized by the energies flowing through them, giving new life to the adjacent countryside.

To ensure that this ancient fertilization still takes place, the aborigines gather at specific locations during certain times of the year, perform the ritualistic dances that have been dictated by the passing time, and pray to the force of the lines. They actually receive messages over vast distances and are forewarned of the approach of strangers – all through their system of magic lines. On the other side of the globe, as recently as the sixteenth century, the Incas used similar spirit lines with the Inca Temple of the Sun in Cuzco as their hub, but no one nation has valued the reported existence of these lines so much as the Chinese. Until the latter part of the nineteenth century, the Chinese practiced an art called *fung-shui*, or 'wind and water', which means 'what cannot be seen and cannot be grasped'. The duty of the practitioners of *fung-shui* was to determine the flow of *lung-mei*, or 'dragon currents', and interpret their influence in the regions they passed through. Every building, stone and planted tree was so placed in the Chinese landscape as to conform to the mysterious dragon currents that flowed along the lines. The main paths of the forces, the Chinese believed, were determined by the routes of the sun, moon and five major planets across the heavens. The dragon currents controlled Chinese life to a great extent, and in the feudal days of China, the Emperor would emphasize the country's dependence on the mysterious force by climbing an artificial mound called Col Hill near Peking several times a year to measure both the planetary and terrestrial energies and to fuse the two for the benefit of the land. Some researchers contend that this was an attempt to combine magic with reality.

The ancient legends of earth currents which are affected by planetary motions and which in turn affect fertility are not fantasy or religious superstition; they are based on very real scientific principles. We are only beginning to realize today that the entire surface of the earth is bathed in the energy of the earth's magnetic field and that this field is subject to certain influences from above and below. The strength and direction of the magnetic currents vary according to the positions of the sun, moon and the closer planets in much the same way as the ebb and flow of the tides take place according to the position of the moon. At the same time, characteristics of magnetic currents are also influenced by the terrain over which they flow. A flat landscape exhibits placid and regular activity, while rocky or broken land shows disturbed behaviour. Magnetic flows are especially agitated over the geologic faults over which many of the prehistoric ley lines have been found.

While some investigators are studying the variations of current on the earth's surface, others are attempting to discover what effects these same currents have on certain nonliving and living components. After exhaustive research, involving 200,000 experiments over ten years, Giorgio Piccardi, Director of the Institute for Physical Chemistry in Florence, Italy, has concluded that water is extremely sensitive to electromagnetic fields, and that as the fields are changed or influenced, so the chemistry of water may be altered. Piccardi also found that since the earth's energy field is subject to change by changes in the positions of the sun and moon, chemical reactions using water as a base also change accordingly. The Florentine chemist's work has been verified by W. H. Fisher of the National Centre for Atmospheric Research of Boulder, Colorado, who noted further that since water is the liquid of life, electromagnetic fluctuations could thus affect growth. Drs A. A. Boe and D. K. Salunkhe, horticulturists at Utah State University, have come up with significant results. When green tomatoes, for example, were placed within a magnetic field, they ripened four to six times faster than under normal conditions. The researchers also noted that seeds of a variety of plants grew many times faster than usual when they had been placed in a current. More recent research has uncovered not only that the living plant is stimulated, but that currents also affect the soil in which the plant grows. The movement of galactic bodies, it now appears, causes certain magnetic fluctuations which in turn increase the fertility of plants as the chemistry of the mineral content of the soil is changed.

Of course we are only beginning to understand the principles behind celestial and terrestrial magnetic influences on the earth's surface, but it seems that people in centuries past not only knew these principles but applied them for their own benefit. First, these early people must have had the scientific knowledge to be aware of the currents' existence, so they developed a technology for detecting those currents. Second, they must have possessed the end product of a tremendous body of research and experimentation that covered centuries, perhaps, and by which they knew how to manipulate the currents towards a predictable result. We can only guess at this from what remains of the earlier culture, for we ourselves have not yet reached that level in our understanding. It appears that the currents began at certain natural energy 'springs' in the earth, which were later marked as religious sites, and from here the currents were directed to specific centres – towers or mounds – where they were gathered and from which they were eventually dissipated to the surrounding country-side. Astronomical observation was of paramount importance, for only by a constant watch over the celestial movements could the waxing and waning of the currents be measured and anticipated.

Directing the magnetic currents seems to have been accomplished by the placement of the standing stones along the ley lines. Throughout recorded history, special powers have been attributed to many of these stones by local traditions. A dolmen or stone group near Finistre, France, for example, is said to cure rheumatism; other neighbouring stones are believed to heal fever and paralysis. Modern investigators have discovered that mysterious energies do emanate from a number of the stones, for photographs of them are sometimes marred by a mist of light surrounding their lower extremities. According to Gey Underweed, author of *The Pattern of the Past*, the standing stones served the same purpose as the needles of Chinese acupuncture. Just as the needles are claimed to redirect the flow of 'life forces' in the human body to restore health, so the standing stones were placed in such a manner so as to realign earth magnetism from the natural paths to artificial ones. Using special dowsing equipment for detection, Underwood found evidence that the magnetic currents in parts of his native Britain do in fact run in rows parallel to the straight lines of the standing stones with a precision that characterizes human construction rather than natural patterns.

Ancient legends explain that the major purpose of the ley-line systems was for the increase of soil fertility and plant growth, but there appear to have been other uses for the lines as well. The modern Druids claim their forefathers built the ley lines and were able to utilize the linear energies for flight. On the day a line became 'animated' by a sunrise directly down a path, the currents were directed so as to charge a body to such a degree that it could be levitated and made to move along the path of a specific level of magnetic intensity. Druidic tradition tells of such heroes as Mog Ruith, Bladud and the magician Abiris, who possessed flying vehicles activated by the ley-line energies and were able to travel in them as far as Greece. The stories of these flights usually end in disaster – an eclipse takes place, suddenly terminating the power sources along the lines, and the hero and his craft plunge to earth and destruction. There is scientific fact behind these stories, for an eclipse of the sun or moon does indeed cause a sudden drop in the level of the earth's surface magnetic activity. We find that among many ancient cultures around the world there was an obsession with predicting eclipses. Perhaps the reason was not so much superstitious fear, as has always been assumed, but changes that eclipses wrought upon terrestrial magnetic flows.

Looking at what we now know concerning the ley lines, we

can make several observations. Examples of ley-line systems are found all over the earth – in Europe, Africa, Asia, Australia, and the Americas – and the legends and traditions associated with them indicate that they were all based on the same principle, magnetic manipulation, and were used for the same purposes. Their conception certainly did not originate with one group of isolated people and then slowly spread to other neighbouring groups; rather, the system appears to have sprung up all over the world simultaneously, planned by a culture that had surveyed the globe and charted the geographical features that revealed the underlying major centres of magnetic activity and inactivity. The very nature of the operation of the lines required that for the system to work to its full potential, all the terrestrial surface currents had to be accounted for. The ley-line system was thus a truly global system. Commenting on the lines, John Michell writes in *The View over Atlantis*, 'A great scientific instrument lies sprawled over the entire surface of the globe. At some period – perhaps it was about 4000 years ago – almost every corner of the world was visited by a group of men who came with a particular task to accomplish. With the help of some remarkable power, by which they could cut and raise enormous blocks of stone, these men erected vast astronomical instruments, circles of erect pillars, pyramids, underground tunnels, cyclopean alignments, whose course from horizon to horizon was marked by stones, mounds, and earthworks.'[4]

Such a global undertaking implies the existence of a single authority directing a unified effort involving the inhabitants of the whole world. Also, just as local sections of the ley lines had a specific centre or even several nodes where the energies converged, so it is likely that the single authority operated from a world centre where the energies of the entire global line system were gathered. The system appears to have operated for a period of time, but then something happened – something significant enough to mark a break in world conditions and to bring the world line system to an end. Before the event, the construction of the system had necessitated a unified world. At some specific point in time that unity was decisively broken. The single directing authority lost its power, and its world centre ceased to operate. Following the event, new conditions prevailed, and the people of the world were fragmented into factions, making unity of effort and the co-ordinated working of the ley lines no longer possible.

As Michell described it, 'All we can suppose is that some overwhelming disaster, whether or not of natural origin, *destroyed a system whose maintenance depended upon its control of certain natural forces across the entire earth.* All attempts at reconstructing whatever it was that collapsed during the great upheaval have ever since been frustrated by schism and degeneration. Falling ever deeper into ignorance, increasingly at the mercy of rival idealists, the isolated groups of survivors all over the world forgot their former unity, and, in the course of striving to re-create some local version of the old universal system, perverted the tradition and lost its spiritual invocation.'[5]

Eventually, even the perversions – the myths and legends of past powers – were partially garbled or forgotten, and the surviving local systems were abandoned. Today we are left with only shadows and remnants of the former universal system.

Thanks to the discoveries made by archaeologists, much of what was considered legendary in the Genesis account of the rise of nations is now being found to be true. The story of a former world unity which was broken into factions is beginning to take on a realistic form. Founded on even older historical accounts, the Genesis story of the Tower of Babel relates the desperate efforts of the new generations to remain together 'lest we be scattered abroad upon the face of the whole earth.' So they began to construct a world centre and a tower that was to reach to the skies.

It appears that as the post-Flood populations in the settled land around the globe were growing large enough to form the bases of active cultures, the people of the world were worried that their continued unity might one day be dissolved. Their fears were justified, for they were trying to re-establish the one-world civilization that the antediluvians had possessed. Babel was chosen to become the capital of the world, symbolizing the organization of the post-Flood peoples under a centralized authority, in the same way that Cain had organized his descendants under one rule by means of the construction of Enoch City. The city of Babel represented a 'United Nations', or a political centre for world government. The Tower of Babel, on the other hand, intended to be a great structure reaching to the skies, may have represented something even more significant. As noted earlier, there very likely had been a world centre where the surface energies of the globe were eventually gathered from the global ley-line system. We know that the place where the currents were accumulated was usually characterized by a mound or tower. The Tower of Babel may have been the receiving station for the ley-line currents of the earth, By their possession of such a centre of the world's energies, the ruling authorities at Babel literally controlled the world, for everyone who desired to benefit from the world ley-line system would have had to serve the rulers of Babel.

We know from all accounts that the lines were used for occult purposes, so there were spiritual as well as material energies involved. The post-Flood ley-line system was very probably a reconstruction of a system used before the Flood. The antediluvians had developed a sophisticated form of technology that incorporated the use of both material and occult energies as its power base, and the ley-line system was simply a further extension of this occult technology.

CHAPTER 4

Advanced Aviation in Prehistoric Times

Soon after the destruction of the Babel world centre, a number of secondary civilization centres emerged in various parts of the world. The initial catastrophe that cast the globe into a period of chaos and confusion may have lasted as long as a century, during which time many of the pre-Babel nations lost contact with one another, while still others were probably overrun by migrating tribes uprooted in the confusion.

The technology of many of them was not affected, however, and they were able to maintain their high level of sophistication and knowledge. The Renaissance maps tell us that at least five generations of ancient cartographers from a highly advanced civilization made a series of uninterrupted surveys of the world before, during and at the end of the Ice Age. The conditions under which they operated must have been subject to major modifications, for whereas in pre-Babel days they had lived under a 'one world command', the post-Babel world presented an entirely different situation. The world was now split into political factions, each one claiming sovereignty over and independence of the others. The former global co-operation had evaporated, and the various political and national entities began to strive for world domination. During the first part of the century of confusion after Babel, the rivalry did nothing to disturb the balance of power, but towards the end of the first hundred years, natural catastrophes, probably initiated by the beginning of the Ice Age, greatly affected the already uneasy truce that existed between them. The violence that followed led to the mutual destruction of these groups. The advanced technology they had struggled so hard to preserve for their national greatness now spawned an arsenal of weaponry that ultimately destroyed them.

There were eight centres of post-Babel high civilization where the remnants of the pre-Flood technology were treasured and utilized. These centres were in the Middle East, northern Europe, the Arctic, India, what is now the Gobi Desert, Antarctica, western and central South America, and south-western North America and the Caribbean. During the initial period of chaos, these centres were cut off from one another, but communications were soon re-established. One of these contact methods appears to have been by air, strange as that may seem in our age, for we still consider the Wright brothers' invention a spontaneous outburst of creativity befitting the twentieth century. Many legends recorded by subsequent peoples contain remembrances of a period when aviation was a well-known concept and flight was a frequent occurrence.

One of the earliest preserved records of flight is in a Babylonian set of laws called the *Halkatha*, which contains this passage: 'To operate a flying machine is a great privilege. Knowledge of flying is most ancient, a gift of the gods of old for saving lives.'

The Babylonian 'Epic of Etana', describing a prehistoric flight, is preserved for us in fragmentary text and cuneiform dating back to a period between 3000 and 2400 BC. The epic tells of Etana, a poor shepherd who finds an eagle with injured wings. He nurses the eagle back to health, and in return the eagle promises to take him on a journey up into the heavens. Etana thereupon mounts the bird, and together they soar off into the sky, gazing at intervals down on the earth below.

At the first stage, the eagle cries, 'Behold, my friend, the land and how it is! Look upon the sea also. Lo, the land has become like a hill and the sea like a watercourse!' This observation is made after an ascent of a double hour's march – in modern terms, six to eight miles high. Rising high above Mesopotamia, Etana can see the mountains of Armenia in the north, and to the south-east, the 'sea' of the Persian Gulf looking like a 'watercourse' or river stretching to the horizon.

According to the inscriptions, the two climb three double marches higher before the eagle again directs Etana's attention to the earth's appearance. From here, he says, the earth looks like a 'plantation' and the land is like a 'hut' surrounded by the 'courtyard of the sea'. Etana has reached a height from which he can see the waters of the Indian Ocean, the Red Sea, the Mediterranean and the Black Sea, encompassing the Middle East.

As they look from still further up, the land appears to be like a 'grinding stone' and the sea like a 'gardener's canal' or irrigation ditch. The mountains are no longer distinguishable; from this altitude the surface appears even and curved, like the surface of a grindstone. At the circumference Etana sees the waters of the oceans surrounding Asia, Europe, and Africa.

Higher still, the eagle remarks that the earth now looks like a 'garden', the sea like a 'wicker basket'. The various characteristic shapes of the continents are now apparent, as are their orange deserts, dark green forested areas, grey valleys and brown-yellow mountains, which look like the coloured patches of a garden. Now Etana also sees the oceans of the world, this time no longer as a surrounding ditch but as separate basins, like various 'baskets' filled with water.

Finally he reaches a height where he is unable to distinguish between land and sea. He has reached an altitude where the clouds and water vapour of the atmosphere hide most of the earth's features in a bluish-white haze. At this point the journey ends, and Etana is returned to earth.

The only myth element the epic contains is the eagle, who may represent some form of aircraft that with the passing of time was transformed into a bird by a people ignorant of the mechanics of flight. Whatever the vehicle of ascent may have been, the 'Epic of Etana' certainly supplies us with a very accurate description of the earth's surface from various altitudes – descriptions which were not verified in our own era until the high-altitude aerial flights of the 1950s and the first space shots of the 1960s. Question: Who made and recorded this observation in the ancient East before 2400 BC?

Another Chaldean work, the *Sifr'ala*, dates back more than 5000 years, and though fragmentary, it is a work filling almost a hundred pages of English translation. Archaeologist and ethnologist Y. N. Iban A'haraon, who worked on its decipherment, found to his astonishment that the *Sifr' ala* is a detailed account of how to build and operate an aircraft. The text speaks of various parts such as vibrating spheres, graphite rods and copper coils; and on the subject of flight the writer comments on wind resistance, gliding and stability. Unfortunately, many key lines of the text are missing, making any attempt at reconstructing the craft impossible.

Early Chinese annals also contain several references to the art of flying. Emperor Shun, who reigned between 2258 and 2208 BC, reportedly not only constructed a flying craft but also tested a parachute – more than thirty-six centuries before Leonardo da Vinci.

In 1766 BC another Chinese Emperor, Cheng Tang, ordered a court artisan named Ki-Kung-Shi to construct a flying apparatus. The craftsman built the machine and flew it to the province of Honan on a test flight. The Emperor, however, had the aerial craft destroyed so that its secret might not fall into the wrong hands.

The secret of flight seems to have survived until the third century BC, for the Chinese poet Chu Yun penned his experiences while flying in a jade-coloured craft over the Gobi Desert to the Kunlun mountains in the south-west. He made an aerial survey of the region and accurately described how the high-soaring craft was unaffected by the wind and dust of the wasteland below.

And even as late as the fourth century AD another Chinese writer, Ko-Hung, spoke of a 'flying car' made of wood and possessing 'rotating blades' that caused the car to travel skyward. In the same century, a flying craft also appeared in Ceylon, where the Buddhist monk Gunarvarman used it to fly to the island of Java – a distance of 2000 miles.

References to flight also appear in the *Budhasvamin Brihat Katha Shlokasamgraha* of Nepal, a twelfth-century written version of an oral tradition of unknown age. It was first published in Europe in Felix Lacote's French translation in 1908. The *Brihat Katha* tells the story of Rumanvit, the servant of a king who desired to travel about the earth in a flying vehicle. In order to satisfy his master, Rumanvit commanded the court designers to construct the needed flying apparatus, but they informed him that they were unable to do so. They knew the workings of many machines, they declared, but the secret of flying machines was known only to the 'Yavanas'.

Yavana was the Sanskrit name for the lighter-skinned peoples of the eastern Mediterranean. More specifically, *Yavana* is derived from Javan, the name of one of Noah's grandsons, whose descendants inhabited mainland Greece and the Mediterranean islands in the first few centuries following the Flood.

The story of Rumanvit ends with the appearance at his master's court of a Yavana from the west who fulfills the monarch's wish to see the world from the air, but without revealing to him the mechanics of flying. There appears to have been a conscious effort on the part of the high civilization centres not to proliferate advanced technology among those post-Babel peoples who had lost knowledge, but rather to keep that technology for their own use and power.

Ancient Flight in the Pacific

Legends very similar to those of the Nepalese are found among the Polynesians. On the South Pacific island of Ponape, the natives tell of learned men with lighter skins than their own, who came from the west long before the European explorers arrived. These former light-skinned men came in 'shining boats' that 'flew above the sea'. Their stay was very brief, but the natives still speak of the 'magical works' the ancient Westerners performed.

The aboriginal inhabitants of Mangareva, the largest of the Gambier Islands, also have a tradition of flight which dates from the ancient past. They recount how a 'flying canoe' with 'great wings clasped tightly to the side' appeared before them, and how the 'priests' who operated it were able to fly great distances – as far as the Hawaiian Islands, nearly 2500 miles away. Robert Lee Eskridge, a collector of Polynesian folklore, found a native on the island of Tara-Vai who gave him a detailed description and showed him an actual artist's model of the ancient flying canoe. According to Eskridge, it certainly represented some form of flying apparatus, and the wings in particular reminded him of those of the winged solar disc of the god Horus, frequently pictured in Egyptian art.

The Saqqara 'Bird'

In 1898, a small model plane was discovered in a tomb near Saqqara, Egypt, and was dated at approximately 200 BC. At the time of its discovery, the birth of modern aviation was still several years away, and so, when the strange object was sent to the Cairo Museum of Antiquities, it was catalogued as Special Register No. 6347, Rm. 22, and then shelved to gather dust among other miscellaneous artifacts – unrecognized for what it really was.

In 1969, Dr Kalil Messiha, an Egyptologist and archaeologist, was cleaning out the museum's basement storage area when he happened on a box marked simply 'bird objects' and discovered the model. The other contents of the box were obvious bird figurines, but one artifact was definitely out of place, possessing characteristics which, though not found in birds, yet are part of modern aircraft. Dr Messiha, who as a youth had been a model-plane enthusiast, immediately recognized the aircraft features and persuaded the Under Secretary of the Egyptian Ministry of Culture, Dr Mohammed Gamal El Din Moukhtar, to form a committee to investigate the model. The research committee was formed on 23 December 1971, and consisted of a number of historians and aviation experts. They were so impressed by the preliminary findings that they recommended the model be hung as a centrepiece in the Central Hall of the Cairo Museum.

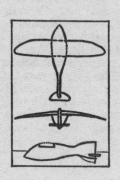

An artist's conception of the Saqqara "bird" as it was discovered in an Egyptian tomb in 1898. This object, which at first was thought to be a model of a bird, flies perfectly as a glider, even though there are indications that it may originally have possessed a propulsion mechanism at the tail. The design of the "bird" is highly sophisticated.

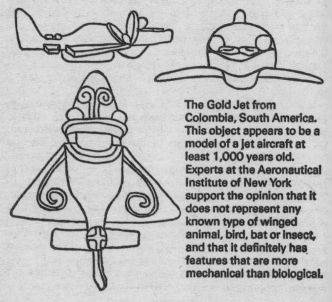

The Gold Jet from Colombia, South America. This object appears to be a model of a jet aircraft at least 1,000 years old. Experts at the Aeronautical Institute of New York support the opinion that it does not represent any known type of winged animal, bird, bat or insect, and that it definitely has features that are more mechanical than biological.

The model's wings are straight and aerodynamically shaped, with a span of 7·2 inches. The pointed nose is 1·3 inches long, and the body of the craft measures 5·6 inches long, tapered and terminating in a vertical tail fin. A separate slotted piece on the tail is precisely like the back stabilizer section of a modern plane. The small craft is made of very light sycamore wood and weighs 1·11 ounces.

When asked to analyze the model, several aerodynamics engineers and pilots found a number of remarkable features, all indicating knowledge of principles of aircraft design which had taken European and American designers a century of airfoil experimental work to discover and perfect. Besides an aerodynamic shape of fuselage and wing that revealed design compensation for camber – the rise of the curve of the wing – the wing itself was found to be counter-dihedral, which provided a tremendous lift force. It appears the ancient craft's purpose was more for carrying large amounts of freight than for reaching high speeds, for designers agreed it could have carried heavy loads, but at extremely slow speed, i.e., below sixty miles an hour. One expert, in fact, noted that there is a remarkable similarity between the down-pointing nose and pointed wing of the Egyptian plane and a new oblique-wing aircraft under consideration by NASA. It too is specifically designed for heavy cargo and low-powered flight. We do not know, however, what the power source of the ancient craft was. The lower part of the tail is jagged – evidently something has been broken off – so that it may have held some form of motor. The engineers noted that the model did make a perfect glider just as it was; in fact, it would have taken only a small catapult to get a life-sized model into the air. Even today, though it is over 2000 years old, the little plane still soars through the air for a considerable distance with only a slight push of the hand!

Another feature aerodynamics experts discovered when they attempted to make a blueprint of the plane was that all of its highly accurate integral proportions were present in ratios of 2:1 or 3:1. It is clear that the ancient model was not accidental or meant only to be a toy; rather, it was the end product of an enormous body of computation and experimentation. Dr Messiha noted that the ancient Egyptians always built scale models of everything they made, for their tombs were filled with small detailed temples, obelisks, houses, chariots, ships, etc. Now that a model plane has been found, Dr Messiha wonders whether somewhere under the desert sands along the Nile there may be the remains of life-sized gliders.

More recently, several other model planes have been uncovered from other tombs and identified, bringing the total number of Egyptian gliders to fourteen. As biologist-zoologist Ivan T. Sanderson, head of the Society for the Investigation of the Unexplained, commented, 'The concrete evidence that the ancients knew of flight was forced upon us only a few years ago. Now we have to explain it. And when we do we will have to rearrange a great many of our concepts of ancient history.'

A Gold Plane from the New World

In 1954 the government of Colombia sent part of its collection of ancient gold artifacts on a tour to six museums in the United States. During the US tour, Emanuel M. Staubs, one of the leading jewellers in America, was commissioned to make cast reproductions of six of the gold pieces. Fifteen years later, one of the casts was given to Ivan T. Sanderson for an analysis. After making a thorough examination of the artifact and consulting a number of aerodynamics experts, Sanderson came to a mind-boggling conclusion. In his opinion, the gold object is a model of a jet aircraft at least a thousand years old.

The object is approximately two inches long and was worn as a pendant on a chain around the neck. Discovered in northern Colombia, the artifact has been classified as Sinu, a pre-Inca culture dating between AD 500 and 800. For want of better identification, the Colombian government labelled the find a 'zoomorfica'; that is, an animal-shaped object. From a zoological standpoint, however, both biologist Sanderson and Dr Arthur Poyslee of the Aeronautical Institute of New York concluded that the object does not represent any known type of winged animal, whether bird, bat, insect, flying fish, skate or ray. In fact, the little Colombian artifact has features that are more mechanical than biological.

Among the important features are the front wings, which are deltoid, with perfectly straight edges – very animal-like. Aircraft designer Arthur Young also noted that if the gold object did represent a flying animal, the front wings are located in the wrong place. They are too far back on the body to coincide with the animal's centre of gravity. The wings are in the right place aerodynamically, however, for a tail-engine jet.

Test pilot and aerodynamics expert Jack A. Ullrich pointed

out further that the delta shape of the front wings and the aero-dynamic tapering of the fuselage imply that the original aircraft was jet-powered, with the ability to fly at supersonic speeds.

After examining close-up photographs of the gold model, taken from the front angle, still another aircraft engineer, Adolph Heuer, noted a third indication of the original plane's potential of preformance. While most modern planes have wings angled slightly upward, only the higher-powered planes have wings that tilt downward. This feature can be seen on the supersonic Concorde, and it can also be seen on the Colombian gold object.

The tail is perhaps the least animal- but most airplane-like feature of the gold model. It is right-triangular in shape, flat-surfaced, and rigidly perpendicular to the body and delta wings. No bird or insect has a tail like this. Only fish have upright tail fins, but none has an upright fin without a counterbalancing lower one. The triangular configuration of the gold model, however, is standard design on modern aircraft.

Another interesting feature about the tail is the insignia that appears on the left face of the rudder, precisely where identifi-cation marks appear on many airplanes today. The insignia is perhaps as out of place as the gold model itself, because it has been identified as the Aramaic or early Hebrew letter beth, or B. This would indicate that the original aircraft may have come not from Colombia but from somewhere in the Middle East.

The gold airplane is by no means the only such model aircraft discovered in the New World. Six very similar gold objects, each complete with aerodynamically designed fuselage, wings, and right-triangular rudder, are on display in Chicago's Field Museum of Natural History, and two others are on exhibit in the Smith-sonian Museum of Natural History in Washington, DC, and in the Museum of Primitive Art in New York City. Along with those in Bogota, Colombia, such objects number fourteen in all. Again, they are well over 1000 years old, but the area from which they come is quite extensive. These other planes were discovered in Costa Rica, Venezuela, and Peru. If people from the Middle East did make flights across the Atlantic, they must have made a number of contacts with the semiprimitive inhabitants in both Central and South America. Looking at the models together, we find they appear to be variations of a single aircraft design. They are either an artist's impression of something he saw himself or his interpretation of a mythical or legendary description of aircraft from the more distant past. The early form of the

Hebrew beth on the Colombian model strongly supports this conclusion and puts the original aircraft and its flight to the Americas before the second millennium BC.

The Hindu Vimanas

Some of the most remarkable descriptions of prehistoric aircraft come to us from India. Among the ancient Hindu sacred books we find the *Samaranga Sutradhara*, a collection of texts compiled in the eleventh century but which date back to unknown antiquity. The *Samaranga* contains 230 stanzas that describe in detail every possible aspect of flying, from how the apparatus was powered to the proper clothing and diet of the pilots. Recently the International Academy of Sanskrit Research in Mysore, India, conducted a special study of the ancient work and published its findings in a book entitled *Aeronautics, a Manuscript From the Prehistoric Past*. The text revealed a knowledge of aircraft design, function and performance that is above and beyond what the laws of chance would permit had the work been only the product of someone's imagination. The following are a few translated excerpts from the text:

'The aircraft which can go by its own force like a bird – on the earth or water or through the air – is called a Vimana. That which can travel in the sky from place to place is called a Vimana by the sages of old.'

'The body must be strong and durable and built of light wood [Laghu-daru], shaped like a bird in flight with wings outstretched [mahavinhanga]. Within it must be placed the mercury engine, with its heating apparatus made of iron underneath.

'In the larger craft [Daru-vimana], because it is built heavier [alaghu], four strong containers of mercury must be built into the interior. When these are heated by controlled fire from the iron containers, the Vimana possesses thunder power through the mercury. The iron engine must have properly welded joints to be filled with mercury, and when fire is conducted to the upper part, it develops power with the roar of a lion. By means of the energy latent in mercury, the driving whirlwind is set in motion, and the traveller sitting inside the Vimana may travel in the air, to such a distance as to look like a pearl in the sky.'

Conspicuously missing from the ancient text is any distinct description of how the Vimanas were actually constructed. The reason for the lack of detail, the ancient sages declared, was that 'any person not initiated in the art of building machines

114

of flight will cause mischief.' In other words, the intricate knowledge of aircraft and flying in the post-Flood era was carefully controlled by a select few.

The chief puzzle concerning the Hindu Vimanas as they are described in the *Samaranga*, however, is their propulsion, which as the text stated was somehow supplied by 'the energy latent in mercury'. It is interesting that the element mercury had a special place in the sciences of the ancients and of the alchemists of medieval Europe. The British nuclear physicist Edward Neville da Costa Andrade, in a speech delivered at Cambridge in July 1946, noted that the famed discoverer of the laws of gravitation, Sir Isaac Newton, knew something about the secret of mercury. Quoting Lord Atterbury, a contemporary of Newton, Andrade said, 'Modesty teaches us to speak of the ancients with respect, especially when we are not very familiar with their works. Newton, who knew them practically by heart, had the greatest respect for them, and considered them to be men of genius and superior intelligence who had carried their discoveries in every field much further than we today suspect, judging from what remains of their writings. More ancient writings have been lost than have been preserved, and perhaps our new discoveries are of less value than those that we have lost.'

Andrade continued, quoting Newton, '"Because the way by which mercury may be impregnated, it has been thought fit to be concealed by others that have known it, and therefore may possibly be an inlet to something more noble, not to be communicated without immense danger to the world."'

What it is about mercury that could be of 'immense danger' to the world we do not know. Yet it seems apparent that the ancients were well aware of the practical application of mercury. Recently Soviet explorers excavating a cave near Tashkent in the Uzbek SSR discovered a number of conical ceramic pots, each carefully sealed and each containing a single drop of mercury. A description and illustrations of the mysterious pots were published in the Soviet periodical *The Modern Technologist*. There is no clue to what these mercury containers were used for, but they must have been highly treasured and used for something that is beyond our present understanding and technology. It was a secret that was found, used and preserved by a select few – only to be lost again, perhaps forever.

CHAPTER 5

Nuclear Warfare Among the 'Primitives'

The world was the scene of great confusion after the breakdown of unity at the world centre. Maps reproduced by the Renaissance cartographers show that the survey expeditions that roamed the globe recorded ominous changes in the northern and southern polar ice-caps. Every one of the sacred historical manuscripts seems to indicate that Job lived during this time period, for certain areas of his story tell of a significant drop in temperature, of freezing conditions, of ice formations advancing from the north, of lowered ocean levels, of pronounced evaporation and excessive flooding and melting of snow.

He experienced these phenomena while living in Uz, a town located in northern Arabia. Today the climate in this area is extremely arid and hot, yet he alluded to overflowing rivers and to rain and even snow.

Puzzling? No, not to the climatologists, anyway, for to them it is a fact that those weather conditions were once considered normal in the Middle East, during the Ice Age!

The advent of climatic changes and glaciation certainly must have had a dramatic effect on the progressive life-style and development of the civilization centres and may have been instrumental in the destruction of at least three of them. Antarctica was one of the centres that were destroyed by the expanding walls of ice. We will probably never know all the details of the tragedy that struck both civilization and human life in the Antarctic continent. History is frightfully silent on matters relating to the coldest regions of the globe, yet there are several indications from primitive sources that Antarctica was indeed inhabited at a very remote period.

Francis Maziere, who has done extensive research into the legends and folklore of the central Pacific island natives, discovered that the Polynesians possessed a very sophisticated knowledge of navigation and geography. They knew about such far-removed locations as New Zealand, Hawaii, Easter Island, and even the south-west coast of South America, with the treacherous waters of Drake Passage beyond the southern extremity of Cape Horn. The Polynesians were also very familiar with the existence of the Antarctic continent. According to their traditions, there was a time when the land was not covered with ice, and several nations of people inhabited it. The Australian aborigines talk of Antarctica as a 'land of the gods' which at an unknown time became covered with 'cool water and quartz crystals', a good description of ice and snow by natives who had never seen such substances in their native desert home.

Maziere found a Polynesian elder from Easter Island named Veri-veri who related that in the midst of the southern land was a great cliff of red rock. Remarkably enough, an identical landmark was discovered recently by an American expedition that ventured into the heart of Antarctica. The red cliff, however, is situated several hundred miles inland, so that it could not have been observed from the coast. It would have been impossible for a Polynesian to have traversed Antarctica in its present frozen state to see the red cliff and live to tell about it. If an ancestor of the Polynesians did observe the red escarpment, as the legend would indicate someone did, this must have been when the climatic conditions on Antarctica were radically different.

Another area that was overwhelmed by glaciation during the Ice Age was the Arctic region, in particular the island of Greenland. One Renaissance map showing Greenland free of ice is the Zeno brothers' map of 1380.

This map was the result of a voyage made by the two Zeno brothers from Venice in the early fourteenth century. Their explorations supposedly took them to Iceland, Greenland and perhaps as far as Nova Scotia. They drew a map of the North Atlantic which was subsequently lost for two centuries before it was rediscovered by a descendant of the Zenos.

A study of the chart reveals that the Zeno brothers could not have been the original map makers. The brothers supposedly touched land in Iceland and Greenland, yet their chart very accurately shows longitude and latitude not only for these locations, but also for Norway, Sweden, Denmark, the German Baltic coast, Scotland, and even such little-known landfalls as the Shetland and Faroe islands. The map also shows evidence of having been based on a polar projection, which was beyond the abilities of the fourteenth-century geographers. The original map makers likewise knew the correct lengths of degrees of longitude for the entire North Atlantic; thus, it is very possible that the map, instead of being a product after the fact, was drawn up by the Zeno brothers before their voyage and was used to guide them in their exploration of the northern lands.

Just how ancient the original source maps may have been is indicated by the fact that the Zeno map shows Greenland completely free of ice. Mountains in the interior are depicted, and rivers are drawn flowing to the sea, where in many cases glaciers are found today. Captain A. H. Mallery, whose initial work on the Piri Reis map (Chapter 3) led him to study other Renaissance charts such as the Zeno brothers', took special note of the flat plain shown stretching the length of the Greenland interior on this map, intersected midway by mountains. The Paul-Emile Victor French Polar Expedition of 1947–9 found precisely such topography from seismic profiles.

As with the revelation that Antarctica at one time was free of ice and perhaps inhabited, we find similar legends of a civilized people who once lived in northern lands which are now buried under thousands of feet of ice. The legends tell of Thule, Numinor and the Hyperboreans, inhabitants of the Arctic in centuries past. Egerton Sykes, in his *Dictionary of Nonclassical Mythology*, page 20, states his belief that the Norse legend of Fimbelvetr, the 'Terrible Winter' that launched the epic disasters of Ragnarok and the destruction of the gods of Valhalla, may reflect a historical fact: the obliteration of a prehistoric civilization in the boreal regions by the Ice Age catastrophe.

The remains of these civilized northern people have, of course, disintegrated under the weight of millions of tons of moving ice, but some evidence of their historical successors has miraculously survived, for there are extensive ruins of a sophisticated prehistoric culture that once existed in the Arctic region. At Ipiutak on Point Hope, northern Alaska, there are the ruins of a large settlement of 800 structures laid out in carefully planned blocks and avenues – a community large enough to have supported several thousand individuals. Unfortunately there are very few artifacts that can tell us anything about the Ipiutak settlement. What we do know is that the ancient settlement was far from being a simple hunting community There are indications that these people had a knowledge of mathematics and astronomy comparable to that of the ancient Mayas. Archaeologists are astonished that a community the size of Ipiutak could have existed at all, for it is situated on the permafrost, far north of the Arctic Circle, where today small bands of Eskimo hunters scratch out a meagre livelihood. Ipiutak could have supported so large and sophisticated a population only if the climate of Alaska was decidedly different from the present, and the only time when this region was considerably warmer was before at the beginning of the Ice Age.

Ipiutak was very probably settled by those people of the Arctic high civilization centre who escaped the first onslaught of the polar glaciation but were overwhelmed as the freezing conditions advanced further south. The Ipiutak cemetery reveals that the inhabitants were tall, blond individuals, similar to the Cro-Magnons of Europe.

Not long ago Russian archaeologists discovered the remains of a number of prehistoric settlements very similar to Ipiutak in the midst of the frozen taiga in north-eastern Siberia. Here too the climate is very hostile to all forms of life, yet the archaeologists found evidence of large Paleolithic, Neolithic and even Bronze Age populations that appear to have lived simultaneously in the same area. In Yakutia, Paleolithic rock drawnigs have been discovered that are much like the cave paintings of Magdalenian France and Spain. Between Yakutia and western Europe, the land and the prehistoric cultures it supported are completely devoid of evidence of any similar artistic development. The only possible link between Siberia and the European Cro-Magnon civilization is through the north, in the direction of a common homeland and origin in the Arctic. Historian Will Durant, in his *Story of Civilization*, made a statement which may contain more truth than previously realized: 'Immense volumes have been written to expound our knowledge, and conceal our ignorance, of primitive man . . . Primitive cultures were not necessarily the ancestors of our own; for all we know they may be the degenerate remnants of higher cultures that decayed when human leadership moved in in the wake of the ice.'[1]

A third post-Babel high civilization centre to be destroyed by the Ice Age was located in the Caribbean. Since 1968, strange finds have been made in coastal waters around the Caribbean, notably in the Bahama Banks. At depths ranging from 6 to 100 feet there are numerous giant stone constructions – walls, great squares, crosses and other geometric shapes, even archways and pyramids – all encrusted with fossilized shells and petrified mangrove roots, indicating their great age. Among the first finds made were stretches of a wall composed of blocks measuring as much as 18 by 20 by 10 feet and weighing approximately 25 tons each. The wall appears to have encircled the islands of North and South Bimini to form a dike. Along with the sea wall, 3- to 5-foot sections of fluted columns were also discovered, some still fixed in their original positions, while

...ying in a jumble on the sea floor, covered ... the pillars appear at regular intervals along ... wall, it is believed they may have formed one ...us portico. Both the wall and the pillars reveal a high ... of engineering skill in their construction.

Not far from the Bimini sea wall, divers have uncovered a stone archway at a depth of 12 feet, a pyramid with a flattened top and a base 140 by 180 feet, plus a huge circular stone construction, made of 12-foot blocks, that appears to have been a well-designed water reservoir when it existed above sea level.

Andros Island, near Pine Key, possesses its share of submarine structures as well. In 1969 airline pilots photographed a 60 by 100-foot rectangular shape, clearly visible through the calm waters. The eastern side and the western corners were partitioned off. What is amazing is that this submerged rectangle is an almost exact copy in size and design of the Temple of the Turtles, an ancient Mayan sanctuary found at Uxmal in Yucatán, indicating that the survivors of this Caribbean civilization centre may have influenced the development of the early Central American cultures and the culture of the Mound Builders.

Other sunken ruins in the Caribbean area include a sea wall 30 feet high, running in a straight line for miles off Venezuela, near the mouth of the Orinoco River, an acropolitan complex; complete with streets, covering 5 acres in 6 feet of water off the Cuban coast; remains of sunken buildings off Hispaniola, one measuring 240 by 80 feet; several stone causeways, 30 to 100 feet below the surface, which leave the shores of Quintana Roo, Mexico, and Belize, British Honduras, and continue out to sea for miles towards an unknown destination; a sea wall running along a submarine cliff near Cay Lobos; and huge stone squares, rectangles and crosses clearly of human design off the windward and leeward sides of all the keys down to Orange Key.

These Caribbean ruins are perplexing to the archaeologists and to orthodox historians, for the architecture is far beyond the capabilities of either the Amerinds or the Spanish conquistadors. It is even more disturbing that the most recent period, when the present Caribbean sea floor was above sea level and the mystery walls, pyramids and temples therefore could have been built, was during the Ice Age. Apparently the Caribbean civilization evolved during the time the ocean levels were at their lowest, and it eventually was submerged when the Bahama shelf was inundated by the rising of the sea caused by the melting of the northern glaciers. The flooding in all probability was very gradual, for many of the gargantuan submerged walls appear to have been dikes built in an attempt to protect certain areas from the rising sea. But the walls were not enough. The ocean waters eventually rolled over the land, and the Caribbean civilization disappeared.

While three of the post-Babel centres of high civilization succumbed to natural disaster, the ruins of the remaining five centres show evidence of man-made destruction – destruction of such terrifying magnitude that we could not have imagined the like of it prior to the end of World War II.

Evidence of the holocaust is found in the most notable of the Hindu literary works, the *Mahabharata*, an epic poem of 200,000 lines, dating back in its present form to 500 BC. Textual evidence, however, indicates that the events depicted in the *Mahabharata* took place 1000 to 2000 years earlier. Repeated references are made to great god-kings riding about in Vimanas or 'celestial cars', described as 'aerial chariots with sides of iron clad with wings'.

Used for transportation in peaceful times, the Vimanas were also employed during battle. The *Mahabharata* describes an eighteen-day war between the Kauravas and the Pandavas, who inhabited the upper regions of the Ganges. Not long after this war, a second battle was waged against the Vrishnis and Andhakas in the same region. In both battles Vimanas were used to launch a weapon of terrible destructive power. The *Mahabharata* relates, 'The valiant Adwattan, remaining steadfast in his Vimana, landed upon the water and from there unleashed the Agneya weapon, incapable of being resisted by the very gods. Taking careful aim against his foes, the preceptor's son let loose the blazing missile of smokeless fire with tremendous force. Dense arrows of flame, like a great shower, issued forth upon creation, encompassing the enemy. Meteors flashed down from the sky. A thick gloom swiftly settled upon the Pandava hosts. All points of the compass were lost in darkness. Fierce winds began to blow. Clouds roared upward, showering dust and gravel.

'Birds croaked madly, and beasts shuddered from the destruction. The very elements seemed disturbed. The sun seemed to waver in the heavens. The earth shook, scorched by the terrible violent heat of this weapon. Elephants burst into flame and ran to and fro in a frenzy, seeking protection from the terror. Over a vast area, other animals crumpled to the ground and died. The waters boiled, and the creatures residing therein also died. From all points of the compass the arrows of flame rained continuously and fiercely. The missile of Adwattan burst with the power of thunder, and the hostile warriors collapsed like trees burnt in a raging fire. Thousands of war vehicles fell down on all sides.'

The description of the second battle is as frightening as that of the first: 'Gurkha, flying in his swift and powerful Vimana, hurled against the three cities of the Vrishnis and Andhakas a single projectile charged with all the power of the Universe. An incandescent column of smoke and fire, as brilliant as ten thousand suns, rose in all its splendour. It was the unknown weapon, the iron thunderbolt, a gigantic messenger of death which reduced to ashes the entire race of the Vrishnis and Andhakas.

'The corpses were so burnt that they were no longer recognizable. Hair and nails fell out. Pottery broke without cause. Birds, disturbed, circled in the air and were turned white. Foodstuffs were poisoned. To escape, the warriors threw themselves in streams to wash themselves and their equipment. With the destruction ended, the Kuru king, Yudistthira, was informed of the power of the iron thunderbolt and the slaughter of the Vrishnis.'[2]

We could attribute these descriptions to the overactive imagination of some unknown Hindu sage of long ago, but there are too many details that make this unnervingly similar to an eye-witness report of an atomic bomb explosion: the brightness of the blast; the column of rising smoke and fire; the fallout, intense heat and shock waves; the appearance of the victims; and the effects of radiation poisoning.

Hindu scholars believe the ancient atomic explosions occurred in either 3102 or 2449 BC, with the latter as the more probable date, because of the detailed astronomical configuration given in connection with the battles in the *Mahabharata*. If the latter date is correct, this means, in terms of Biblical chronology, that atomic weapons must have been used about a millennium after the Flood. According to traditional Hindu history, the Bharata War took place not many generations after the reign of Manu, who escaped a world-destroying Deluge with his family in a boat – the Hindu equivalent of Noah and the ark.

When European scholars began their first examination of the *Mahabharata* in the nineteenth century during the British imperial rule of India, the many references to flying craft and weapons of fearful fiery destruction were considered nothing more than poetic hyperbole. In the words of one Victorian commentator, V. R. Dikshitar, 'Everything in this literature is imagination and should be summarily dismissed as unreal.' But with the initial research into radiation and nuclear physics at the turn of the century, there were already those who saw in the *Mahabharata* and other ancient legends an indication of energies that were just beginning to be understood by modern man Physicist Frederick Soddy, in his *Interpretation of Radium*, published in 1909, remarked concerning the ancient accounts, 'Can we not read in them some justification for the belief that some former forgotten race of men attained not only to the knowledge we have so recently won, but also to the power that is not yet ours? . . . I believe that there have been civilizations in the past that were familiar with atomic energy, and that by misusing it they were totally destroyed.' Since 1945, of course, we have learned what the effects of the destructive power of the atomic bomb are – and the descriptions given in the *Mahabharata* have suddenly become very real.

The use of atomic weapons in India 4400 years ago presupposes a knowledge of nuclear physics rivalling our own. There is evidence of such a knowledge preserved among the ancient Hindu records. Several Sanskrit books, for example, contain references to divisions of time that cover a very wide range. At one extreme, according to Hindu texts dealing with cosmology, is the *kalpa* or 'Day of Brahma', a period of $4 \cdot 32$ billion years. At the other, as described in the *Bihath Sathaka*, we find reference to the *kashta*, equivalent to three one-hundred-millionths ($0 \cdot 00000003$) of a second. Modern Sanskrit scholars have no idea why such large and such miniscule time divisions were necessary in antiquity. All they know is that they were used in the past, and they are obliged to preserve the tradition.

Time divisions of any kind, however, imply that the duration of something has been measured. The only phenomena in nature that can be measured in billions of years or in millionths of a second are the disintegration rates of radioisotopes. These rates range from those of elements like uranium 238, with a half-life of $4 \cdot 51$ billion years, to subatomic particles such as K mesons and hyperons, with mean half-lives measured in the hundred-millionths, billionths, trillionths, and even smaller fractions of a second. The ranges of ancient Hindu time division and of radio-isotope disintegration thus partially coincide, and the former could have been used to measure the latter.

If the ancient Hindus – or an earlier civilization from which the Hindus inherited their time divisions – had a technology that could study and measure nuclear and subnuclear matter, means for using atomic energy was also accessible to them.

There are remains that strongly suggest that an atomic war was indeed waged in the distant past. According to the *Mahabharata*, the Bharata War, in which Vimanas and atomic weapons were used, involved prehistoric inhabitants along the upper Ganges River in northern India. Precisely in this region, between the Ganges and the mountains of Rajamahal, there are numerous charred ruins which have yet to be explored or excavated. What observations have been made thus far indicate that the ruins were not burned by ordinary fire. In many instances, they appear as huge masses fused together, with deeply pitted surfaces, described as looking like tin struck by a stream of molten steel. Further south, among the dense forests of the Deccan, are more such ruins which may be of earlier origin, pointing to a war antedating that of the *Mahabharata* and encompassing a far greater area. The walls have been glazed, corroded and split by tremendous heat. Within several of the buildings that remain standing, even the surfaces of the stone furniture have been vitrified: melted and then crystallized. No natural burning flame or volcanic eruption could have produced a heat intense enough to cause this phenomenon. Only the heat released through atomic energy could have done this damage.

In the same region as this second group of ruins, Russian researcher A. Gorbovsky reported, in his *Riddles of the Ancient Past*, the discovery of a human skeleton the radioactivity of which was fifty times above the normal level.

Outside India, similar remains of a nuclear holocaust have been found. Researcher Erich von Fange describes the melted ruins of a ziggurat structure situated not far from ancient Babylon: 'It appeared that fire had struck the tower and split it down to the very foundation . . . In different parts of the ruins, immense brown and black masses of brickwork had [been] changed to a vitrified state . . . subjected to some kind of fierce heat, and completely molten. The whole ruin has the appearance of a burnt mountain.'[3]

In 1952 archaeologists excavating in Israel unearthed at the sixteen-foot level a layer of fused green glass a quarter of an inch thick and covering an area of several hundred square feet. It is made of fused quartz sand with green discolourations, similar in appearance to the layers of vitrified sand that were left after the atomic tests in Nevada in the 1950s. Another such sheet of glass was uncovered five years earlier in southern Iraq, near Babylon, spread in a thin layer some distance below Babylonian

124

Sumerian and Neolithic cultural levels. To the south, the western Arabian desert is strewn with black rocks which show evidence of having been subjected to intense radiation. There are twenty-eight fields of these scorched stones, called *harras*, covering an area of 7000 square miles. In the southern Sahara Desert, engineer Albion W. Hart discovered another expanse of green glass and noted that the fused silica there was similar in appearance to that found at the White Sands atomic test site. Still other examples of vitrified soil have been discovered among remains in the most desolate areas of the Gobi Desert of Mongolia. Most surprising of all are layers at Lop Nor in Sinkiang, near the present Chinese atomic test site, where it is reported that there is little difference between the patches of fused quartz left after the modern nuclear detonations and those that had been there ages before the Chinese became a nuclear power.

Elsewhere prehistoric forts and towers in Europe as far north as the British Isles and the Lofoten Islands off Norway have had their walls vitrified and stones fused by an unknown energy, usually along their western sides. Many of the towers of Scotland and the granite fortresses along the coast of Ireland show evidence of having been melted to a depth of one foot.

One of the most amazing literary testimonies to man-made destruction among the ancient advanced civilizations is found in the Tibetan *Stanzas of Dzyan*, translated within the past century, the original dating back several millennia. Like the *Mahabharata*, the *Stanzas of Dzayn* depict a holocaust engulfing two warring nations who utilize flying vehicles and fiery weapons.

'The great King of the Dazzling Face, the chief of all the Yellow-faced, was sad, seeing the evil intentions of the Dark-faced. He sent his air vehicles to all his brother chiefs with pious men within, saying, Prepare, arise, men of the good law, and escape while the land has not yet been overwhelmed by the waters.

'The Lords of the Storm are also approaching. Their war vehicles are nearing the land. One night and two days only shall the Lords of the Dark-faced arrive on this patient land. She is doomed when the waters descend on her. The Lords of the Dark-eyed have prepared their magic Agneyastra [the Hindu "Agneya weapon" – a nuclear missile] . . . They are also versed in Ashtar [the highest magical knowledge]. Come, and use yours.

'Let every Lord of the Dazzling Face ensnare the air vehicle of every Lord of the Dark-faced, lest any of them escape . . .

'The great King fell upon his Dazzling Face and wept. When the kings were assembled, the waters of the earth had already been disturbed. The nations crossed the dry lands. They went beyond the water mark. The kings reached then the safe lands in their air vehicles, and arrived in the lands of fire and metal . . .

'Stars [nuclear missiles?] showered on the lands of the Dark-faced while they slept. The speaking beasts [radios?] remained quiet. The Lords waited for orders, but they came not, for their masters slept. The waters rose and covered the valleys . . . In the high lands there dwelt those who escaped, the men of the yellow faces and of the straight eye.'

Even though the translation of this text was made almost a century ago, it describes forms of destruction we have become familiar with only in the last thirty-two years. It is also significant that the man-made destruction depicted in the text is coupled with cataclysmic movements of ocean waters. This massive flooding may have been touched off by the nuclear holocaust, but more likely the inundation was a result of a sudden sea-level change caused by the melting Ice Age glaciers. If the 'Lords of the Yellow-faced' were the Mongolian inhabitants of the ancient Gobi high civilization centre, the flooding may have been the great tidal wave that swept across eastern Asia and into Siberia at the end of the Ice Age. According to the *Stanzas*, another high civilization centre, described only as the 'Lords of the Dark-faced,' had advance knowledge of the imminent deluge that was about to weaken the Gobi centre, and so decided to take advantage of the situation and destroy the survivors with a nuclear barrage and conventional air attack. The 'Yellow-faced' seem to have retaliated with a nuclear counter-attack of their own, and while a few of the Yellow-faced escaped the flooding and nuclear destruction, the Dark-faced and their civilization appear to have been annihilated. The final line of the text mentions that among the survivors were also those of 'the straight eye', the peoples of Eurppe and the Middle East. This would suggest that these people also were involved in the nuclear conflict; the remains of nuclear destruction in these areas bear testimony to this.

One of the most interesting pieces of evidence of nuclear devastation in the past is found on Easter Island, in the Pacific Ocean. Apart from its huge monolithic statues and its curious form of writing, the island is also famous for a unique form of wood carving called *moaikavakava*. The carving invariably represents a shrunken man, with certain grotesque anatomical features depicted in remarkable detail. The first Europeans to visit Easter Island reported that the natives were often willing to part with these little statues, as if the figures did not belong to them. Even today the miniature men are regarded by the Easter Islanders as fearful and alien – a reminder of something that was not of their experience, yet which remains horrifying none the less.

Native legend attributes these statues to King Tu'ukoiho. One night the king caught a glimpse of two misshapen dwarfish beings who he believed were the spirits of the last members of a race that had inhabited the island before the present native population. Even though they were never seen again after his one fleeting glance, the impression these wretched men made on the king was so strong that he immediately set out to sculpt a replica of them. The modern *kavakava* statues are believed to be faithful copies of the king's original.

The style of these carvings is not in the least Polynesian, and the sculpted facial features – hooked nose, staring eyes and small squared beard – appear to be Semitic. The most interesting peculiarity, however, is the appearance of the body. It is emaciated, showing goiters, tumours, clenched mouth, collapsed cervical vertebrae, and a distinct break between the lumbar and the dorsal vertebrae. All these are medical indications of exposure to a severe dosage of radiation.

Perhaps related to the *kavakava* carvings and the unfortunate victims they represent are the remains of fiery destruction discovered on the island. At the foot of the slopes of Mount Rano-Kao there exists an immense furrow, half a mile in length and approximately 200 yards wide. The furrow is sharply defined on the landscape, because it is composed of obsidian, a vitrified black rock which is not found anywhere else on the island. Directly aligned with this trail of melted rock is a small crater on a hill a mile away. The crater is perfectly circular and is distinguished by a vegetation cover different from that growing around it. The furrow and crater imply that in the unrecorded past something landed here with tremendous force. Whether that something was natural or man-made, of course, is open to speculation. The *kavakava* statues and their indication of intense irradiation of the island at some time in prehistory, however, might possibly favour the latter possibility.

The continents of the New World also possess several examples of prehistoric cultures destroyed by a great conflagration. Not far from Cuzco, Peru, near the pre-Inca fortress of Sacsahuaman, an area of 18,000 square yards of mountain rock has been fused and crystallized. Not only the mountainside, but a number of the dressed granite blocks of the fortress itself show signs of similar vitrification through extremely high radiated heat.

In Brazil there is a series of ruins called Sete Ciddaes, situated south of Teresina between Piripiri and the Rio Longe. The stones of these ruins have been melted by apocalyptic energies, and squashed between the layers of rock protrude bits of rusting metal that leave streaks like the traces of red tears down the crystallized wall surface.

The most numerous vitrified remains in the New World are located in the western United States. In 1850 the American explorer Captain Ives William Walker was the first to view some of these ruins, situated in Death Valley. He discovered a city about a mile long, with the lines of the streets and the positions of the buildings still visible. At the centre he found a huge rock, between 20 and 30 feet high, with the remains of an enormous structure atop it. The southern side of both the rock and the building was melted and vitrified. Walker assumed that a volcano had been responsible for this phenomenon, but there is no volcano in the area. In addition, tectonic heat could not have caused such a liquefaction of the rock surface.

An associate of Captain Walker who followed up his initial exploration commented, 'The whole region between the rivers Gila and San Juan is covered with remains. The ruins of cities are to be found there which must be most extensive, and they are burnt out and vitrified in part, full of fused stones and craters caused by fires which were hot enough to liquefy rock or metal. There are paving stones and houses torn with monstrous cracks . . . [as though they had] been attacked by a giant's fire-plough.'

Other vitrified ruins have been found in parts of Southern California, Arizona and Colorado. The Mohave Desert is reported to contain several circular patches of fused glass.

If an unknown post-Flood civilization was indeed destroyed by fire in western North America, we would expect that such a holocaust would have been imprinted on the memory of those who survived, to be told and repeated to the successive generations. While studying Canadian Indian tribal folklore, ethnologist R. Baker was told the following legend by a wise man of a dying totemic cult of northern Canada near the tundra region. The legend tells of a time 'before the cold descended from the north,' when the now-bleak tundra was instead rich in vegetation.

'In the days when great forests and flowering meadows were here, demons came and made slaves of our people and sent the young to die among the rocks and below the ground [mining?].

128

But then arrived the thunderbird, and our people were freed. We learned about the marvellous cities of the thunderbird, which were beyond the big lakes and rivers to the south.

'Many of our people left us and saw these shining cities and witnessed the grand homes and the mystery of men who flew upon the skies. But then the demons returned, and there was terrible destruction. Those of our people who had gone southward returned to declare that all life in the cities was gone – nothing but silence remained.'

This is all the totemic Indians know of the matter, and they cannot furnish any additional details about these events. This is the story that has been repeated to them by their fathers and forefathers.

The Hopi Indians of the South-west have a very similar tradition which offers yet another glimpse of otherwise unrecorded events. The story is called 'Kuskurza', the 'Third World Epoch', and is preserved in Frank Waters's *Book of the Hopi*.

'Some of these of the Third World made a *patuwvota* and with their magical powers made it soar through the sky. On this many of them flew to a great city, attacked it and returned so quickly that the inhabitants did not know where their attackers came from. Soon others from many nations were making *patuwvotas*, and [they] flew to attack one another. So corruption and destruction came to the Third World people, as it had come to those who were before.'

The ruins in the western United States show signs of having been destroyed by radiated heat, mute testimony to an element of fact that may underlie these legends. In addition, the association of the annihilation of prehistoric cities with men who flew through the air is disturbingly similar to the Hindu and Tibetan records of air vehicles armed with nuclear weapons.

With the collapse of the post-Babel centres between 2900 and 2800 BC, the world once again entered a brief period of confusion and adjustment. To those inhabitants of the high civilization centres who managed to survive, two choices were left. Either they could begin again by re-establishing their own cultures, or they could migrate and become members of those inferior cultures, the Stone Age civilizations or the emerging civilizations of the Mediterranean and Middle East, areas which for the most part were unaffected by the natural and man-made catastrophes. Of those who chose the first alternative, many survived only as simple farmers. Richard Mooney comments in his book *Colony Earth*: 'The collapse of a technologically superior civilization would have left little time for salvaging anything but essentials. The survivors may have salvaged certain needed devices, among them a few aircraft, which would have been required to maintain contact with other survivors. In the course of time, it would have become more and more difficult to keep such machines in working order. Parts would wear out, sources of fuel and power would fail. If the technology is destroyed, the capacity for manufacturing machine parts or even making the right metals will no longer exist . . . In the end, all that would be left would be a memory of strange dart-shaped flying things in which people once travelled through the sky. Ages later, who would believe such fantastic stories?[4]

When Noah and his family escaped the destruction of the antediluvian civilization, they had passed through 120 years of preparation, during which they had time to gather the knowledge necessary to live and to begin a new culture in the post-Flood era. But the destruction of the post-Babel centres must have come swiftly and without warning, allowing no time to preserve much of the knowledge. With the extinction of the technological environment – the technical resources, and the co-ordination and specialization of labour by both men and machines – the degree of civilization that survived must have been very limited indeed. Without industry, the survivors had to concentrate all their efforts on producing their own essentials, and the establishment of agricultural self-sufficiency must have been the first priority. The tragedy is that once the knowledge possessed by the high civilization diminished, the descendants of these surviving farmers remained farmers and nothing more.

It is now evident that the survivors' offspring did in fact establish a number of important agricultural centres. Up until two decades ago, archaeologists were certain that agriculture was first practiced in the Middle East's 'fertile crescent', and that from there it eventually spread throughout the entire world. But new excavations in various sections of the globe have considerably changed this erroneous concept. The new evidence now points to the existence of major agricultural centres in north-eastern China, south-eastern Asia, Mexico and Peru that are just as old as that in the Middle East. These findings have caused noticeable consternation among orthodox historians, for they provide more questions than answers. The most disturbing question of all is why these agricultural centres suddenly appeared in different parts of the world at approximately the same time. A probable answer may be found in their locations. Middle East agriculture

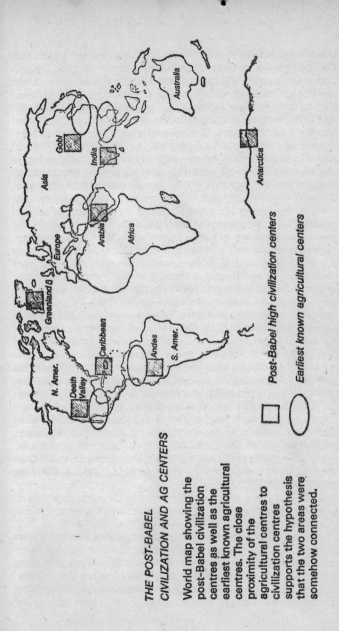

THE POST-BABEL
CIVILIZATION AND AG CENTERS

World map showing the
post-Babel civilization
centres as well as the
earliest known agricultural
centres. The close
proximity of the
agricultural centres to
civilization centres
supports the hypothesis
that the two areas were
somehow connected.

□ Post-Babel high civilization centers

○ Earliest known agricultural centers

blossomed not far from those areas in present-day Israel, Iraq and Arabia which were destroyed by nuclear fire; the Chinese and Asian agricultural centres are only a short distance from the Gobi and Indian civilization centres, respectively; the Mexican agricultural sites are just south of the Death Valley ruins; and the Peruvian agricultural centre is in the same locale as the melted façade of Sacsahuaman. More recently, an additional agricultural centre was discovered in Venezuela, not far from the vitrified ruins in the Brazilian jungle. Because almost total destruction hit the centres almost simultaneously, it is only reasonable to assume that the survivors would have developed their individual farming communities within a short time of each other. The same development can be seen in the history of pottery. Historians thought for many years that the Middle East was the home of the world's first pottery industry, just as it had been for agriculture. Since then, however, pottery as old as the earliest examples in the Middle East has been found in Japan. Anthropologist J. Edmonson constructed a theoretical framework in which he attempted to trace the first pottery in the Middle East and Japan, as well as later finds in Asia and Africa, back to a common origin. The theoretical centre he discovered was Ulan Bator, in Mongolia, precisely in the middle of the former Gobi high civilization site!

The second choice open to the remnants of the races was to migrate to and share what knowledge they still possessed with the inferior civilizations of Europe and the Middle East that were unaffected by the nuclear and Ice Age catastrophes. The impact of a higher civilization on a lower one, of course, would have produced profound results. This is exactly what we find to be true in the histories of the known cultures. Paleolithic civilization in Europe did not originate from Europe, but came in successive influxes from the north and west. Similarly, in the myth-histories of the ancient civilizations, we invariably find the following order of events:

1 An initial period of rulership by god-kings of great knowledge [corresponding to the pre-Babel period]
2 A period of confusion and regression, during which primitive cultures briefly flourished [post-Babel]
3 The advent of culture-bearing foreigners and a sudden explosion of architecture, social organization and religion, which remained relatively unchanged in the succeeding millennia.

132

Orthodox historians unfortunately do not recognize the existence of these mythical culture bearers, insisting instead that the ancient civilizations were the result of slow but steady development from Stone Age beginnings. But what these historians have found impossible to explain is why the archaeological evidence points to no transition whatsoever between the ancient civilizations and their primitive forebears. Van der Veer makes the most of this in his book *Hidden Worlds*, where he writes, 'Let us take the Egyptians as an example. The ability to build pyramids demands at the least a knowledge of arithmetic, architectural techniques and skill in transporting materials – all of which suggests a long preliminary period, the existence of which is unfortunately not supported by archaeology. Archaeologists admit there is a problem here, but won't investigate the reasons for it; they simply accept the idea of the other civilizations; little is ever said about the prehistoric background to these civilizations, as a result of which they attempt to condense almost into insignificance a period which must have covered thousands of years between the Stone Age and the dynastic periods.

'It is impossible to accept these arguments; we believe – and perhaps science, too, will come to believe – that at some stage in the very earliest periods of prehistory, contact was made between the ancient peoples and a still older race in possession of an advanced civilization and a history stretching a long way back indeed. It may be that there is a grain of truth in all mythological stories and legends, and that somewhere on our planet there once existed a race with a very sophisticated civilization which perished because of one or more natural disasters. The only really satisfactory theory is that the survivors of this civilization were responsible for both the technical skills and the art of writing possessed by the old cultures, who brought knowledge to the people then living in the Stone Age.

'What we think happened was this. Somewhere on earth a civilization arose, or perhaps several civilizations, which because of extremely favourable conditions flowered earlier than any others . . . We believe that one or more natural disasters destroyed this original civilization, and that its survivors found refuge among the inhabitants of those favoured areas which later gave rise to the ancient civilizations we now know about. This is the only theory which virtually meets all the objections and fills even the gaps we are reminded of in old myths and legends.'[5]

CHAPTER 6

Unravelling the Enigma of the Cave Man

As the eight world civilization centres destroyed one another, the suffering planet throbbed with pain and terror. Everywhere death rained from the skies. Dense arrows of flame and mushrooming clouds of fire unleashed by the Agneya weapon spewed radiating waves of death over the battlefields, vapourizing both men and machines. The knowledge that had been so carefully preserved and carried through the Flood now became the tool of destruction. Death ruled, and its horrifying stench of decay hung heavy where once proud cities had stood. Gone was the global unity – confusion was rampant.

With knowledge fractured, communications nonexistent, and distrust and hatred the common denominator among the warring nations, ideas and concepts could no longer be exchanged, and the flow of inventiveness and technical advancement abruptly ceased. It was as if a giant hand had suddenly demolished the nations, grabbed the strings of knowledge and pulled them back.

The world was to be changed for a second time. Abandoning the nightmare of the molten cities, leaving them in the clutches of atomic radiation, small groups of panic-stricken survivors set out to begin life once again in the mountains and jungles which were untouched by the holocaust. Finding refuge in caves and crevices marked the beginning of a new existence, far different from the dubious blessings society had brought them. And while the crumbled civilizations sought ways to re-establish themselves, the people trying to recall from memory what had once been entrusted to scrolls and metal plates, the 'cave men' isolated themselves from the mainstream. Their remains are still found today, contributing to the conflict called evolution.

The new framework of history, based on discovery and manuscript translation covering the activities of the human race since the Flood, indicates that there really was no progressive succession. Instead the developments of the Stone Age and the cultures of Egypt and Mesopotamia were merely discontinuous offshoots of the world fragmentation after the building of the Babel World

Centre. Limited (primitive) and advanced civilizations existed at the same time, with each one aware of the others' existence.

Death of the 'Ape-Man'

To prove their theory, scientists up until a few years ago were classifying various prehistoric human skeletal remains into various positions on a hypothetical line of ascent, beginning with the so-called ape-man and ending with modern man. More recent finds, however, have revealed the disconcerting fact that the basic human has always existed, not as the offspring of apes or primitive beings, but as a man, since time began. Those known to us as ape-men were simply humans who had degenerated from the main human stock. Bjorn Kurten, author of *Not from the Apes*, says, 'It has been possible in the last decade to demonstrate that the human lineage can be followed back into far more distant times where it still retains its unique character. Indeed, we may doubt that our ancestor was ever what could properly be called an ape. This makes excellent sense zoologically. The contrasts between apes and men in anatomy . . . are too great to be reconciled with a relatively recent common origin, and the same is true of behaviour.'[1]

This is truly the age of discovery, even though not everyone agrees with the conclusions reached. The 'evolution' of man, as seen through his technological regression, indicates that man did not evolve; rather, he regressed. For nearly a century, Neanderthal man, whose partial skeletal remains have been discovered throughout Europe, was thought by the evolutionists to have been a direct ancestor of modern man. But more recent Neanderthal finds in the Middle East are more advanced, almost like *Homo sapiens* in appearance, yet they are older than those found in western Europe, forcing the palaeontologists to concede that the West European Neanderthalers constituted a step backward. The most satisfactory explanation for the degeneracy of the European Neanderthalers is as follows. By their own volition the people severed their contacts with the civilization centres, and they presently found themselves cut off from the rest of mankind by the Ice Age glaciers that blanketed northern and central Europe. Because of this isolation and their limited numbers, considerable inbreeding occurred. With such a limited gene pool, the appearance of bad genetic traits was significantly increased, leading to birth defects and physical mutations which produced the structures characteristic of West European Neanderthal remains.

There are some palaeontologists who are already beginning to believe that this explanation may apply not only to Neanderthal man but to the rest of the primitive 'ape-men' as well. Harold G. Coffin, Research Professor of the Geoscience Research Institute in Berrien Springs, Michigan, comments: 'Neanderthal man and Cro-Magnon man are not a very useful support for evolution, for they are so much like modern human beings. This is especially true since the recent discovery that the classic descriptions of Neanderthal man were based in large part on the remains of a Neanderthal skeleton of a man suffering from severe osteoarthritis.'

An article entitled 'Pathology and the Posture of the Neanderthal Man', by researchers William L. Straus, Jr., and A. J. A. Cove, lends considerable weight to this evaluation: 'There is thus no valid reason for the assumption that the posture of Neanderthal man of the fourth glacial period differed significantly from that of present-day man,' they point out. 'This is not to deny that his limbs, as well as his skull, exhibit distinctive features – features which collectively distinguish him from all groups of modern men . . . It may be that the arthritic "old man" of La Chapelle-aux-Saints, the postural prototype of Neanderthal man, did actually stand and walk with something of a pathological kyphosis; but, if so, he has his counterparts in modern men similarly afflicted with spinal osteoarthritis. He cannot, in view of his manifest pathology, be used to provide us with a reliable picture of a healthy, normal Neanderthalian. Notwithstanding, if he could be reincarnated and placed in a New York subway – provided that he were bathed, shaved, and dressed in modern clothing – it is doubtful whether he would attract any more attention than some of its other denizens.'[2]

There are already some palaeontologists who are beginning to believe that this explanation, as well as that of recessive genetic traits, may apply not only to Neanderthal man but to the rest of the primitive 'ape-men' as well. Two defects associated with recessive genetic traits are endocrine and thyroid disorders affecting the development of bones and other tissue, and resulting in acromegaly and cretinism. The medical descriptions of these two disorders are similar to the modern palaeontological descriptions of 'ape-men' remains.

Such conditions occur rarely among populations with wide ranges of breeding choice, but, as mentioned above, they can become predominant in a people closely in red because of isolation. With this in mind, it is interesting to note in what locations the remains of major prehistoric 'ape-men' types have been found:

- Pithecanthropus was located in *Indonesia*
- Sinanthropus in east *China*
- Australopithecus in *South Africa*
- Most primitive Neanderthalers in the *western part of Europe*

When we look at these localities in terms of the population dispersal from Ararat following the Flood, we see that Ararat constituted a central starting point, and the primitive men's remains are found on the outer fringes.

While the designation 'Stone Age' clearly does not apply to these remnants of a chaotic culture who carved out a meagre existence far from the mainstream of civilization, it will have to suffice for want of a more appropriate description. The remains of these survivors are usually found in close proximity to the materials that were most durable, stone or bone, hence this name. Yet this does not preclude their having worked with these materials exclusively; in fact, there is evidence that they, like their more civilized neighbours, not only knew about, but worked with metals. It is true that no actual metal tools have ever been discovered among Stone Age relics, but this is understandable, as metal tools will not last much longer than a few thousand years when exposed to the weathering processes of time. That the survivors indeed knew of the value of metal becomes evident when we consider the many prehistoric mines that have been located throughout the world. On the Mediterranean island of Elba, there are iron-ore mines whose origins are lost in antiquity. The Greeks considered the mines already ancient in their own day and ascribed their origin to the Pelasgians, a prehistoric people who inhabited the eastern Mediterranean region.

Beyond Europe, a number of recently excavated sites have greatly increased our knowledge of prehistoric mining operations. Investigations conducted in 1967 and 1969 at Lion Cavern, near Ngwenya in Swaziland, southern Africa, have shown that long before the present Negroid population of Bantus, Bushmen and Hottentots inhabited the area, at a time when local Neanderthal types such as Rhodesian, Boskop and Florisbad man were extinct, someone had already mined deposits of hematite and specularite, forms of iron ore. This hematite has been found in conjunction with Neanderthal remains at La Chapelle-aux-Saints in France and dates back to the same period as the

Ngwenya mines. It is now believed that the 'bloodstone' (hematite) was used as a cosmetic and also for ritual purposes as a substitute for human blood in burial ceremonies. The use of hematite in this specific manner has been discovered as far away as Tasmania, off southern Australia, and Tierra del Fuego, at the southern tip of South America – *always in coastal areas*. It is possible that the use of bloodstone, and perhaps the material itself, may have been exported over a considerable area in prehistoric times. This extensive trade, of course, was totally out of keeping with modern theories of the primitiveness of early man.

Not far from Ngwenya, at Border Cave in South Africa, diggings in 1972 conducted by Adrian Boshier and Peter Beaumont uncovered ten filled-in prehistoric mining pits, some up to forty-five feet in depth. Again, hematite had been extracted. Associated with the Border Cave remains were remains of both Neanderthal and modern types of primitive man. Also found were agate knives still sharp enough to cut paper, as well as evidence that the miners used mathematics and kept records by making etchings on bone. It would appear that the ore had sufficient economic value to prompt the primitive diggers to keep track of what they produced.

Interestingly, some of the most fascinating evidence of pre-historic mining is found in North America. In the Keweenaw Peninsula and on Isle Royale in Michigan, in the copper-rich Lake Superior region, there are ancient mines whose origins are completely unknown even to the Indians. There are signs that several thousand tons of copper were removed at a very early date, yet not a single cultural artifact remains that could tell us who the miners were. The *American Antiquarian* (vol. 25, p. 258) remarks, 'There is no indication of any permanent settlement near these mines. Not a vestige of a dwelling, nor a skeleton, nor a bone has ever been found.' What is known is that the prehistoric miners had the means not only of extracting the ore, but also of transporting it to a distant location, for not one ounce of the ore was ever uncovered for use within a thousand miles of the mine sites.

The first discovery of the prehistoric mining shafts was made in 1848 by S. O. Knapp, an agent of the Minnesota Mining Company. In passing over a portion of the company's grounds, he observed a continuous depression in the soil, which he surmized was formed by the disintegration of a vein. The depressions led him to a cavern, where he noticed evidence of artificial excavation. After clearing away the debris, he discovered numerous stone hammers, and at the bottom of the hole was a vein of ore which the ancient miners evidently had not finished unearthing.

Two and a half miles east of the Ontonagon River, today the centre of the copper region of Michigan, Knapp discovered a second mine. This shaft was situated in a rock wall. The excavation, which reached a depth of twenty-six feet, had later been filled in with clay and a tangled mass of vegetation – indicative of an extremely old mine. At a depth of eighteen feet, Knapp uncovered a detached mass of copper weighing six tons. This mass had been raised on timbers and wedges to about five feet above its break-off point. The timbers were from six to eight inches in diameter, and the ends showed the marks of a cutting tool. The copper mass itself had been pounded smooth, and what had been protruding pieces were broken off to facilitate transportation. The shaft contained other copper masses, charcoal and other evidence of fire, and a stone hammer weighing thirty-six pounds.

On Isle Royale, near the northern shore of Lake Superior, prehistoric excavations are extensive, with some pits reaching sixty feet in depth. Upon opening one of the island pits, searchers discovered that the mine had been worked to a depth of nine feet through solid rock before a vein of copper eighteen inches thick was uncovered at the bottom. Obviously the miners were highly intelligent and experienced both in the observation of locating the veins and then in following them underground when their course on the surface was interrupted. Many of the excavations were connected underground, and drains were cut into the rock to remove excess water. At one point, the Isle Royale excavations extend for two miles in an almost straight line.

Still more curious than the Michigan copper mines is this find reported in the February 1954 issue of *Coal Age*. During the preceding year, miners at the Lion Coal Mine in Wattis, Utah, broke into a pre-existing tunnel system, of which there was no modern record. The tunnels were so old, in fact, that the coal residue in them had already oxidized to a great extent and could no longer be of commercial value. On 13 August 1953, John E. Wilson of the Department of Engineering and Jesse D. Jennings of the Department of Anthropology of the University of Utah began an exploration of the prehistoric coal mines They found not only tunnels, but also centralized coal rooms where the material had been brought before being transported to the surface.

The tunnels averaged five to six feet in height and extended for several hundred feet, following seams of coal in patterns similar to modern mine layouts. The scientists were unable to find the surface entrance of the old mine system, but they did trace an eight-foot-high tunnel to a depth of 8500 feet. Subsequent investigation revealed that no local Indian tribe had ever used coal or had a recollection of anyone who did. As with the Michigan mines, some enterprising prehistoric people not only had possessed the technology for mining the ore, but also had the means of transporting the material to some unknown location.

Construction Techniques of the Stone-Age Men

A lack of metal artifacts certainly does not prove that the people of the Stone Age did not use metals, nor does the fact that most Paleolithic remains have been found in caves mean that they were the single mode of habitation among Stone Age men. Le-Grand-Pressigny in France has the most extensive deposits of stone tools in the world – millions of cores and scrapers from the Paleolithic Age are scattered over 10,000 acres, at depths averaging three to six feet – yet there is not a single cave in the area. At Charroux is another tool centre of considerable size, where even today one can pick up prehistoric stone axes over twenty-five acres. Within three miles of the Charroux site, in the hillsides along the Charente River, there are forty-nine caves, but excavations have revealed no sign that any of these caverns was ever inhabited by men.

Evidence that Stone Age men lived in well-constructed houses is slowly surfacing and has upset preconceived views of how they lived and flourished. In the Lascaux Caverns, world-renowned for their Magdalenian paintings, one can still see the holes in the rock that supported wooden crossbeams. Probably looking similar to what Michelangelo utilized many millennia later, these crossbeams held scaffolding that enabled the Cro-Magnon artists to execute their works on the cave ceilings, ten to twelve feet above the cavern floor. The evidence for this scaffolding is significant, for in the opinion of Professor Doru Todericiu of the University of Bucharest, the history of architecture shows that scaffolding did not precede knowledge of masonry. If the Lascaux artists constructed scaffolds, it is probable that they also knew how to construct walls. 'To deny this,' Professor Todericiu states, 'would be like saying that the candle was invented before anyone knew how to kindle fire.'

Several examples of simple prehistoric stone construction have been found which show a remarkable degree of sophistication. The Abbé Breuil and Professor Lantier, in their book *Les hommes de l'age de la pierre ancienne*, discuss the finding of a prehistoric oven at Noailles: '[It was] made of squared stones held in place by a packing of chalky clay and sand.' In other words, the Stone Age oven had been constructed using stones shaped like bricks and mortared with cement.

Even in eastern Europe, where the early inhabitants did not share the higher culture of the Magdalenian people of France, we also find indications of a sophisticated knowledge of construction. The remnants of three huts of that period were recently excavated at Vestonice on the lower slopes of the Pavlov hills in Czechoslovakia. The largest of the three was thirty by forty feet in size, and its floor had been covered with limestone grit, a crude form of cement. The smaller huts had been built in similar fashion, using circular walls covered with limestone and clay. These are considered to be among the oldest true walls surviving in the world. What is also significant about the Vestonice site is that a well-constructed beehive-shaped kiln containing remnants of fired clay was found in one of the huts. Fragments of sculptured clay heads of a fox and two bears were also unearthed. Thus the use of fired clay was not beyond the scope of Paleolithic culture, as had previously been thought.

What are perhaps the most disturbing prehistoric construction and civilization finds were uncovered in 1965 by archaeologist Dragoslav Srejovic at a site now called Starveco, on the Danube River, on the Yugoslavian and Rumanian border. Digging into the Yugoslavian bank, Srejovic first encountered traces of a Roman road; beneath this were fragments of proto-Greek pottery, and below these were Neolithic remnants and traces of Mesolithic cultural artifacts. Deeper still, Srejovic came upon something totally out of place: the remains of a cement floor. More specifically, the material was an amalgam of local limestone, sand and water, considered a feat of chemistry and construction several millenia ahead of its time. The cement surfaces were not placed haphazardly, but were carefully laid out in large slabs to form the foundations of houses. Several foundations were built one on top of another, indicating that buildings had been constructed and reconstructed over an indeterminate period. Yet there was also remarkable uniformity. The layout of the houses in the later periods was the same as that in the earlier periods – there was no evidence of a gradual development from a simple to a complex pattern. Rather, the Starveco village suddenly appeared, fully mature, flourished, then decayed and was abandoned in the same advanced state.

In addition to the foundations, the individual Starveco buildings also showed a high order of architectural sophistication. They all had one side larger in size than the other three, with proportions of either 3:1 or 4:1. The larger side was shaped like a sixty-degree segment of a circle. This larger side always faced towards the river, providing the occupants with the maximum view of the Danube and the surrounding hilly country. Inside each house, the shape of the dwelling was repeated in the hearth or oven, which was bounded by carefully shaped stone slabs and always located in the eastern or sunny end of the house. Srejovic noted that the position of the hearth was significant, as it was situated in the exact centre of an equilateral triangle if the lines of the house were extended. What the architectural purpose of this was is not clear, but the implications of the mathematical and geometrical knowledge indicated cannot be ignored.

The same precision and order evident in the architecture is also found in the arrangement of the dwellings at the Starveco site. The structures are laid out in what appears to have been a planned fan shape, opening towards the riverbank. The larger buildings, presumably those belonging to members of a higher class or governing body, were located towards the centre, surrounding a paved plaza believed by Srejovic to have been a market-place or assembly square.

The Starveco site has yielded a number of other cultural characteristics previously thought to have been developed thousands of years later, in the Middle East. Behind the hearth in each house, labourers unearthed the remains of altars, indicating religious beliefs and practices. Each altar was composed of a flat stone, with a cup impression for burning a sacrifice, which faced two or more upright stones of reddish sandstone. This sandstone had been excavated from an outcrop, located in a ravine several miles away, and many of the stones had carved wavy lines or chevrons in low relief, considered the oldest examples of architectural decoration. Even more significant was the discovery of twenty sculpted life-sized human faces of stone. The faces were goggle-eyed, open-mouthed and small-nosed, with some of the statues showing a suggestion of shoulders, arms and a bust. The Starveco sculptures are believed to be the oldest such life-sized, hand-fashioned stone works known today.

An interesting aspect of the site was the evidence of very good health among the Starveco population. There was a striking absence of deformed or diseased bones, and the women were so robustly built that it was difficult to tell their skeletal remains from those of the men. Both sexes lived unusually long lives – some into their eighties. This was indeed an increase over the lifespan of those who inhabited the region during the later Neolithic, Greek and Roman periods, when fifty years of age was considered old.

Community Life and Trade

Among both the cavern and constructed habitations that existed during the Stone Age, we find ample evidence that the inhabitants brought with them concepts of community co-operation and communication. At Les Eyzies, in the Dordogne region of France, numerous caves and rock shelters are clustered together; all were inhabited at the same time. Evidence of co-operation among the cave dwellers begins early, with the Aurignacian period, when the region was occupied by just a few individuals. Larger hearths indicated not only an increased population, but also more complex social units. Similar kinds of tools were found together, indicative of a specialization in both labour and the sites of labour. A number of the sites were used only occasionally, and the tools and bones uncovered were associated with hunting spring and summer game. Ideas were also shared from site to site. Several caves possessed drainage ditches, running through the floor to the outside; all were of similar design and construction. Ideas and concepts must also have been shared over an extensive area, for among many of the Les Eyzies caverns are fragments of seashells, indicating contact with a coastal region – 100 miles away.

Other indications show that the cave men had an intimate knowledge of the seas and must have been familiar with sea travel. As previously noted, the cave-man civilization first appeared along the western coasts of France and Spain, from the direction of the sea. A bone baton found at Montgaudier is engraved with figures of a spouting sperm whale and two seals so detailed that they can be recognized as male and female. Montgaudier is over 100 miles from the coast, indicating that someone knowledgeable about marine life had recorded his observations, which record had found its way far inland from its source. Similarly, in the cave of Nerja, in the Malaga region of southern Spain near the Mediterranean coast, at a deep and nearly inaccessible place on the cavern wall are painted three dolphins, two males and one female, in a face-to-face encounter. Their creator – like the person who carved the image of the sperm whale – would have had to journey far out on the open sea in order to witness and record his story.

If they did voyage by sea, how far did they travel? Evidence of their journeys has been found in coastal areas throughout the western Mediterranean – in Tunisia, Sicily, Italy, Morocco and southern Spain. Even further away, Aurignacian tools and skeletal remains have been uncovered *in the New World*. Professor J. L. Myers, in the *Cambridge Ancient History* (vol. 1, p. 48) noted conspicuous similarities between Aurignacian skulls found in Europe and prehistoric skulls uncovered in Lagoa Santa in Brazil and other localities along the coast of eastern South America. Van der Veer reports that obsidian tools from El Ingor, in the Andes mountains near Quito, Ecuador, are definitely related in design to tools belonging to the late Upper Paleolithic in France and Spain. Stone Age man must have had a considerable knowledge of geography and navigation in order to reach and trade with these distant locations.

Sophistication in Clothing

When the average person imagines a man of the Stone Age, he usually pictures a crude-looking individual, dressed only in an animal skin around his waist and over one shoulder. For decades this was how anthropologists viewed prehistoric man. However, in a cave near Lussac-les-Chateaux, in 1937, Leon Pericard and Stephane Lwoff uncovered a number of engraved stones dating from the Magdalenian period which drastically altered the accepted picture. The flat stones showed men and women in casual poses, wearing robes, boots, belts, coats and hats. One engraving is a profile of a young lady who appears to be sitting and watching something. She is dressed in a pant suit with a short-sleeved jacket, a pair of small boots, and a decorated hat that flops down over her right ear and touches her shoulder. Resting on her lap is a square, flat object with a flap that folds down the front, very much like a modern purse. Other examples show men wearing well-tailored pants and coats, broad belts with clasps, and clipped beards and moustaches.

The Lussac etchings contradict everything that classical prehistory had believed until that time, and anthropologists

An artist's sketch of an etching found in a cave near Lussac-les-Chateaux in France, showing the profile of a young lady who appears to be wearing modern-type clothing, including a hat, and carrying a purse — out of place in the Magdalenian period of the Upper Paleolithic age.

were quick to label the drawings a fraud. But despite their hasty judgements, the out-of-place pictures were authenticated in 1938, with Abbé Breuil among those who demonstrated that the well-dressed individuals had indeed lived during the Magdalenian period of the Upper Paleolithic. Today, most of the stone engravings are in the prehistory library of Lussac-les-Chateaux, with a few on exhibit in the Msuée de l'Homme in Paris. But the drawings that are shown are those which are not too revealing and do not clash too strongly with conventional theories. The rest are stored away and cannot be seen, except by special per-

ission, and then only by those individuals with 'proper creden-
als'. It is felt that the pictures would be too 'disturbing' for
public viewing.

The Lussac models are by no means the only evidence of
sophisticated dress from the Stone Age. Prehistoric cave paintings
from the Kalahari Desert and South-west Africa, dated within
the Stone Age period, show light-skinned men with blond beards
and well-styled hair, wearing boots, tight-fitting pants, multi-
coloured shirts, and coats and gloves. Further to the north, the
remains of a Paleolithic man were uncovered near Vladimir,
not far from Moscow, by Professor Otto Bader of the Ethno-
graphical Institute of the Academy of Soviet Sciences. Christened
'Vladimir man', the prehistoric individual was a hunter of
reindeer and mammoth, and the remnants of his clothing
indicated he was well attired. He wore a large pair of trousers
made of fur, an embroidered shirt, and a very practical jacket.
Scarcely anything remains of the actual clothing, but the pieces
could be reconstructed from the ivory badges and clasps that
were still intact.

Primitive Art – Far Ahead of Its Time

Without question, the most universally recognized aspect of
Stone Age civilization is its art work, which has come to us in a
variety of forms, the most awe-inspiring being the polychrome
paintings found in the caverns of Lascaux, Altamira and other
caves in southern France and northern Spain. Paleolithic art
first appeared with the advent of Cro-Magnon in the Aurignacian
period and became more pronounced and widespread in the
Gravettian. Sculptured female statuettes, currently called the
Venus figurines, are found associated with most Gravettian
remains from France, across Europe and Asia, as far east as
Siberia. But it was not until the Magdalenian period, which must
truly be called a Renaissance, that art burst forth in a wide range
of styles and media of art.

The Magdalenian cave art and the way it was created tell us
much about the sophistication of their culture. The first step in
the execution of a cave painting was to sketch the animal or
other subject matter in outline. This was done either in charcoal
or by engraving with a flint. Following this came the application
of colour, which was accomplished in a variety of ways: with the
fingers; with brushes of fur, feathers or mottled twigs; with pads
of moss; by blowing dry colours through a hollow reed or bone

146

tube; or by rubbing on the colours after mixing them with animal fat and rolling them into crayons. A number of these crayons were found at Altamira.

The colours the cave man had available were somewhat limited. He did not use blues or greens, but utilized a violet-black pigment made with manganese oxides. Chemical analysis shows that the most commonly used colours were yellow (from ochre, i.e., iron oxides), red and orange (from iron oxides and bison's blood), and brown and black (from heated animal fat and charcoal).

The artists achieved a remarkable three-dimensional effect by utilizing the natural contours of rock on the cavern walls and ceilings. Small holes became the glaring eyes of a bison, cracks became the wounds of a stricken deer, and odd-shaped bulges were incorporated into the painting as the head or back hump of a woolly rhinoceros or mammoth. Even today, as one gazes upon the cavern figures, the contrasts between light and shadow created by the natural rock contours give the impression that the painted animals are alive and breathing, a technique and effect unique in the history of art.

The cave paintings, when closely analyzed, reveal that the sketching and application of colour were done in bold, sure strokes, with few apparent mistakes or corrections. This may suggest that those who executed the art were true masters whose confidence and exactness could only have been acquired after years of training and experimentation. At Limeuil, in south-west France, 137 stone slabs were found, with poorly drawn sketches on them, dating from the Paleolithic age. In the midst of each sketch, however, are details redrawn and corrected by someone who was obviously artistically more mature. These drawings show all the signs of a teacher's hand applied to a student's work – a master training the eye of the novice in artistic perception. Limeuil, it now appears, was a school for artists; not only for sketchers, but for painters as well. In an adjacent grotto, a bone tube still filled with paint ready to be blown against the cavern wall was unearthed, as was a stone palette thick with ochre waiting to be applied with a brush.

Not only was art taught, but artistic ideas were conveyed from one place to another, sometimes over great distances. In 1903 a wall picture of a stately old bison, drawn with distinct individuality, was found in a cavern at Font-de-Gaume in the Dordogne, France. Twenty-three years later a stone slate was uncovered in another cave 188 miles away, showing the sketch

147

from which the old bison had been drawn. Someone had undoubtedly admired the Font-de-Gaume painting, acquired its original sketch from the artist, and taken it home to keep as a memento or perhaps to use as a model himself.

In describing the sophistication of Paleolithic art, prehistorian Robert Silverberg says, 'The cave paintings are upsetting to those who prefer to think of Quaternary man as little more than an ape. Not only do they indicate great craftsmanship, but they point to a whole constellation of conclusions: That primitive man had an organized society with continuity and shape, religion and art. It was also dismaying to learn that the earliest inhabitants of Western Europe . . . had scaled heights of artistic achievement that would not be reached again until late in the Christian era. That exploded the theory [that] man's rise from barbarism had been steady and always upward.'[3]

William F. Albright, in his *From the Stone Age to Christianity*, summed up modern research into Paleolithic art in this way: '. . . though the number of motifs, techniques and media available to him now is, of course, immeasurably greater, it is very doubtful whether man's artistic capabilities are actually any higher today than they were in late prehistoric times.'

Examples of Prehistoric Mathematics and Astronomy

Among both cave paintings and various stone and bone engravings we find not only realistic representations of nature and everyday life, but also a great many abstract symbols called tectiforms, claviforms and blazons. Sometimes the forms are recognizable; other times they are not. These symbols no doubt were meant to convey ideas and thus may be considered a Stone Age form of pictograph writing. In many instances, these abstract signs are simply composed of a series of lines, scratches or dots in carefully planned patterns. At first many prehistorians regarded the series only as crude forms of decoration, but now they are identified as notation – some strictly mathematical, others of a chronological nature, recording such astronomical phenomena as the phases of the moon.

One of the most intriguing specimens of prehistoric notation was found on a mammoth tusk from Gontzi, a late Paleolithic site west of Kiev in the Ukraine. The notation appears around the edges of a flattened surface, marked off in graduations like the divisions on a modern ruler or slide rule. The markings are grouped along a horizontal line divided into series by longer strokes at specific intervals. There are also a number of symbols or figures appearing along the sequence, pointing to some event at those intervals. Alexander Marshack, an American researcher, analyzed the Gontzi notation and found unmistakable evidence that it was indeed a detailed record of lunar phases. What's more, the notation pointed to its use as a calculator; that is, the phases of the moon could have been predicted in advance. The Gontzi bone was thus a scientific instrument of a high order, demonstrating that Paleolithic man was more than a mathematician and astronomical observer; he was also a scientist who had applied what he had observed, to create a workable formula that reflected the repetition he had seen and measured in the night sky.

Evidence of Contact with Higher Civilizations – The Universal Lunar Calendar

The existence of a lunar calendar used in the Stone Age civilization is significant not only from a scientific viewpoint, but also as evidence of contact between Stone Age peoples and the peoples of the known ancient civilizations. New archaeological research has discovered that almost every one of the ancient cultures of the Middle East and the New World possessed, at the earliest stages of their development, a primarily lunar calendrical system. Professor Richard A. Parker, in a paper concerning the origins of the calendar used by the Egyptian court, notes that in the early dynastic period the system employed was solar and stellar, based on the simultaneous rising of the sun and the star Sirius once a year. Parker also explains that, according to early dynastic symbolism and ritual, there appears to have been an older calendrical tradition which was lunar in character, dating back into predynastic times and to the very beginnings of Egyptian history.

In Mesopotamia, the first calendars of the Sumerian city-states were also lunar. The Sumerian month began with the moon's first crescent, and the lengths of the months varied with the period of the moon, twenty-nine or thirty days – the same breakdown found among the Stone Age recordings. A lunar calendar was also the first calendrical system of the early Hindu and Chinese civilizations. In the Americas, the first Amerind settlers on both the northern and southern continents are known to have had lunar calendars. The Incas, for example, had an official solar calendar; but their division of the year into twelve months hints at an earlier lunar-count tradition.

Historians have argued that the existence of a lunar calendar in the Stone Age and also among the first civilizations demonstrates their succession; that is, the time count of the moon was developed first in the Stone Age, and then supposedly it was gradually transmitted over tens of thousands of years to the first civilized cultures. But the sacred historical manuscripts furnish evidence that instead the Stone Age peoples and the peoples of the ancient civilizations directly inherited a lunar calendar system from a civilization older than them both.

In Genesis 7 and 8 we find the record of Noah's diary of the Flood. The days of the months and the lengths of time Noah gives for the duration of the events signify very little by themselves, but when these are placed in the framework of the present Jewish calendar, we can isolate some rather interesting data. First, ten of the dates Noah records fall in the calendar on the Jewish Sabbath, Saturday. This could not be coincidental, as it confirms that the data were indeed based on a calendrical system similar to the Jewish calendar – a system which has, in fact, remained relatively unchanged in its basic structure for millennia. Noah must have familiarized himself with the intricate apparent movement of the sun, for he also marked off in his diary the passing of a solar year of 365 days. But the most significant fact is that the Jewish calendar, like the calendar which Noah uses, is based on a lunar count of 354 days. This suggests that the lunar calendar had its true origins during the antediluvian period. By Noah's record, we know that the system was in use immediately after the Flood, and no doubt it was transmitted to his descendants. Following the fiasco at Babel, some of these descendants, we know, remained civilized, while others lost their knowledge. But the lunar calendar appears to have been preserved among both the prehistoric primitive men and the post-Babel civilizations.

Out-of-Place Alphabets and Ancient Memories

Perhaps the most significant evidence of contemporary contacts between the Stone Age culture and the Mediterranean civilizations is the discovery of out-of-place writing among Paleolithic remains. A piece of reindeer bone found in a cave near Rochebertier, France, has markings on it that are more than just decoration. They have every appearance of being the letters of some form of writing. At first glance, one might think that this is conclusive evidence of the existence of a written language during the Paleolithic age, but the implications of the reindeer bone go one step further. The letters resemble or in some cases are identical to the enigmatic script of Tartessos, a city civilization that existed in southern Spain and is believed by some to be the Biblical Tarshish. What makes the similarities of the writings truly remarkable is that orthodox prehistorians place the reindeer bone in the Magdalenian period – by their chronology, about 12,000 years old – and the Tartessian civilization recently has been assigned to the period between 2500 and 2000 BC. There is an obvious discrepancy with this dating, for it is highly unlikely that a script, once developed, would have remained relatively unchanged for ten millennia. What the two scripts *do* demonstrate is that the cultures in which they were found must have been contemporaneous, rather than separated by a vast span of time. The date of the peak of civilization in Tartessos is becoming better established, and if there was a contact between the Paleolithic people and the city of Tartessos, then they must have existed in the same time period. Other finds confirm this. Paleolithic antler bones found at Le Mas d'Azil and La Madelaine are inscribed with signs identical to Phoenician script from approximately 2000 BC. Le Mas d'Azil is also the site where many painted pebbles from the Azilian period of the Mesolithic age have been discovered. A number of these pebbles are marked with signs and symbols that were once predominant throughout the Mediterranean – again, between 3000 and 2000 BC.

Among the records and literature of the ancient civilizations are many accounts of the existence of primitive men living and communicating with civilized men in their day. One of the earliest traditions known to historians is the Gilgamesh epic from Mesopotamia, which tells the tale of the hero Gilgamesh and his many adventures in the world immediately after Babel. Gilgamesh's companion in his experiences was a strange individual named Enkidu whose origins are most interesting. As a youth, Enkidu was described as having lived as an animal among the animals. His hair was long, his nails and teeth were developed for gathering and eating herbs, and he was without intelligent speech, precisely as were the more primitive of the degenerate prehistoric types. He was found one day by his civilized contemporaries, who took him captive and taught him the arts of urban living. It is significant to note that Enkidu's background was not unusual. His primitive life seems to have been regarded as an everyday occurrence, implying that other men at that time were known to live under similar conditions. Enkidu's unique role in the story is that he is described as one of those very few 'wild men' who completely adjusted to Sumerian civilization.

In India, another epic story, the *Ramayana*, depicts a race described as 'ape-men' who aided the noble Rama in a war against the Ravana kingdom of Ceylon. The most celebrated of them was their general, Hanuman. His appearance, described in both the *Ramayana* and the *Mahabharata*, is that of an ape, but he was also capable of humour, intelligent speech and great bravery. He was known for his knowledge of the hills and forests (geography) and for his cures from rare plants (herbal medicine). He is represented in India today as a poet who wrote verse on stone. Underlying the legend is a memory of degenerate men who worked in stone. Equally significant is the fact that Hanuman is presently worshipped as a god by millions of devout Hindus living in south-eastern India, in precisely the areas that are richest in Paleolithic remains. As a curious note, many Hindus also believe that the *yeti* – the mysterious 'abominable snowmen' who are thought to inhabit the inaccessible heights of the Himalayas – may be the descendants of Hanuman and his ape-like but intelligent people.

The ancient Chinese likewise described a race of primitive men coexisting with their own civilization, only they were not pictured as a friendly host. The degenerates were called Mao-tse in the Chinese treatise *Shu King* (part 4, ch. 27, p. 291) and are described as 'an ancient and perverted race who in olden days retired to live in rocky caves, and the descendants of whom are still to be found in the vicinity of Canton.' It is interesting to recall that it was in Hong Kong, only a few miles from Canton, that the giant teeth of Giganthropus were discovered. The *Shu King* relates that the Mao-tse once 'troubled the earth, which became full of their robberies.' The Lord Huang-ti, an emperor of the Chinese Divine Dynasty, then saw how these people were without virtue and ordered his generals Tchang and Lhy to exterminate them. Perhaps it was this genocide that accounts for the sudden disappearance of Sinanthropus and Giganthropus from the Chinese palaeontological record.

A remarkably similar description of a race of primitive men is found in the Bible in the Book of Job. The post-Flood patriarch depicted a wild people with whom he did not wish to associate. He described them as living *in solitude* in the wilderness. They ate *grasses and leaves*, often resorted to stealing food, and – like the Mao-tse – were called thieves and robbers. These wild people also inhabited *the rocks* and cliffs and brayed like animals, as they were without intelligent speech. Job condemned them all as 'a scourge to the land' and the 'children of fools'. Many commentators believe that Job was identical to Jobab, the thirteenth son of Joktan, mentioned in the genealogy of Genesis 10. If this identification is valid, it means that Job, a sixth-generation descendant of Noah, lived about 2698 to 2348 BC, which places him and the 'wild people' he described in the immediate post-Babel period.

Elements of Sophisticated Technology in Stone Age Cultures

Not only are there indications of contact between Stone Age cultures and the known ancient civilizations, but we also find instances demonstrating that on occasion prehistoric primitive peoples also communicated with and benefitted from the knowledge of other unknown civilizations of a very advanced order. A number of discoveries suggest the performance of sophisticated surgery in prehistoric times.

Professor Andronik Jagharian, anthropologist, and director of operative surgery at the Erivan Medical Institute in Soviet Armenia, examined a number of skulls from the ancient site of Ishtikunuy, located near Lake Sevan. The site was inhabited by a prehistoric people called the Khurits who settled the area prior to 2000 BC.

Two of the skulls examined by Professor Jagharian revealed extraordinary skill in head surgery. The first is the skull of a woman who died at approximately thirty-five years of age. In her youth she had suffered a head injury which made a hole one-quarter inch in size in her skull. This accident certainly must have left brain tissue exposed, and a considerable amount of blood must have been lost, The prehistoric surgeons skilfully inserted a plug of animal bone, and the woman survived the delicate operation. This could be seen from the woman's skull, as her own cranial bone grew around the plug before she eventually died years later.

The second Khurits skull shows evidence of even more complicated surgery. The skull is of another woman, who was approximately forty years old when she died. A blow to the head had caused a blunt object about an inch in diameter to puncture the skull, splintering the inner layers of cranial bone. The surgeons of 4000 years ago carefully cut a larger hole around the puncture in order to remove the splinters that had penetrated into the brain. Even by modern standards, such an operation would be considered extremely difficult; yet the prehistoric operation was successful. Evidence shows that the woman survived the surgery for fifteen years.

Concerning his examination of both the skulls and the surgical tools found at the Armenian site, Professor Jagharian commented, 'We have found 4000-year-old obsidian razors at Lake Sevan that are so sharp they can still be used today. Considering the ancient tools the doctors had to work with, I would say they were *technically superior* to modern-day surgeons.'[4]

Evidence of sophisticated prehistoric surgery believed to be even older than the Khurits finds of Armenia was uncovered in 1969, when a Russian expedition of researchers from the universities of Leningrad and Ashkhabad, led by Professor Leonid Marmajarjan, discovered thirty skeletons in a cave in central Asia. Dating techniques placed the age of the remains within the early Paleolithic period. The skeletons were moved to the University of Ashkhabad, where an extensive scientific examination was undertaken.

In a report given to the Soviet Academy of Sciences in November 1969, it was noted that a number of the central Asian skeletons showed signs of surgery having been performed on them. As with the Lake Sevan discoveries, there were several examples of successful operations on the skull. But after examining the skeletons, the Soviet scientists were astonished to find traces of surgery having been performed *in the area of the heart*. The ribs had been expertly cut, and there was also evidence that once an opening had been made, the uncut ribs were further spread apart by retraction. Every feature corresponded to what today is called the 'cardiac window', which enables surgeons to perform open-heart surgery. The periosteum, or bony deposits on the cut ribs, indicated that the patients survived three to five years following this extremely delicate operation.

The success of these prehistoric examples of head and heart surgery testifies to scientific developments which are not only beyond the scope of the Paleolithic and Neolithic cultures as we are beginning to understand them, but also far beyond the developments of most of the ancient and even more recent civilizations. The prehistoric operations presuppose an intimate knowledge of anatomy, especially an understanding of blood flow and its control, as well as advanced notions of hygiene and anesthesia. These points are vital, for without them even the most elementary operation is impossible. Until the last century, the techniques employed in these fields were still so crude that even the amputation of a limb usually resulted in shock or sepsis. What is most significant is that we have as yet found no evidence whatsoever of the development of these advanced medical practices in the Stone Age cultures where the operated skeletons were located. The surgical knowledge must have been borrowed or performed in person by peoples of a highly technical civilization that coexisted with the Stone Age cultures. This is not as incredible as it may seem, when we consider how our present computer civilization is living side by side with primitive Stone Age cultures such as those of New Guineans and the Australian aborigines. And just as modern medical missionaries from our western civilization have saved the lives of thousands of natives in Africa, South America and the Pacific, thousands of years ago unrecognized civilizations utilizing medical knowledge that was just as advanced as ours saved the lives of Stone Age primitives in the same way.

What were the diseases they encountered among their own people and the 'primitives'?

I am sure we will never know exactly the variety of maladies that afflicted early man, but a rare collection of statues in the private collection of Professor Abner Weisman, a New York gynaecologist, has lifted at least part of the ignorance concerning this period.

'When I started my collection in 1944,' Dr Weisman told us, when we first interviewed him for a magazine article a number of years ago, 'most scientists were of the opinion that pre-Colombian art and science were not all that old. Discoveries that have been made in the late 1950s and early '60s have greatly altered that idea. Now we know that several thousand years before the Aztecs, Incas and Mayas, other highly civilized nations occupied that part of America. Their legacy to us did not reach us via a written language, but infiltrated our twentieth century in the form of numerous statues that tell us about the variety of diseases these people suffered.

'What they tell us is simply mind-shattering.'

We gazed at his collection of statues, and suddenly I began to feel sorry for the nation represented by the so-recently unearthed statues. The symptoms of ailments such as cancer, smallpox and osteoarthritis are clearly visible on the often realistically moulded clay statues. Malnutrition, deformities – some of them hideous – pregnancy in various stages, amputations and even birth by Caesarean section are depicted in fine detail. 'Many experts believe that these statues were not really used for

155

instructional purposes, but that they were buried with the deceased to indicate the cause of death. If that is true, then things haven't really changed all that much,' Weisman concluded. 'But it suddenly brings their medical history a lot closer to us.'

One of the most interesting aspects of this collection is that it not only shows the diseases of the ancients, but also supplies hints about the hospitalization of their patients. It is obvious that many of the sick were treated in outdoor facilities, for many of the statues are tied down on rather primitive hospital beds, some equipped with sunshields, while others are on beds where entire sections of the mattresses have been removed, eliminating pressure on bedsores.

In Lima, Peru. Dr Jose Cabrena, professor of anthropology and history at the University of Peru, has collected hundreds of pre-Inca stone carvings discovered in remote areas of the Andes, and these carvings tell of medical knowledge and operating techniques so sophisticated and so refined that our medical scientists of today stand aghast at their implications.

The scenes scraped in ageless rock, made by supposedly ignorant Indians, depict among other things heart transplants, using techniques that seem modern by today's standards. They show Caesarean births, brain transplants, and still other forms of surgery we have developed only within the last generation. Still other stone carvings depict close-ups of heart surgery, showing blood vessels; surgeons at work with their instruments; and patients connected via intricate tubing to life-support systems.

The scientists who have examined the carvings, or photographs of them, are clearly baffled by this discovery.

Dr E. Stanton Maxey, fellow of the American College of Surgeons, says, '. . . in the photographs of stone carvings depicting heart surgery, the detail is clear – the seven blood vessels coming from the heart are faithfully copied.

'The whole thing looks like a cardiac operation, and the surgeons seem to be using techniques that fit with our modern knowledge.

'Another carving shows the surgeons operating on a woman whose full abdomen, enlarged breasts, and what seems to be a foetus strongly suggest a Caesarean-section delivery.

'How such ancient stones can carry a record of modern surgical techniques is completely baffling. It would seem that somehow those ancient people came into contact with a civilization far more advanced than any we have dreamed existed then.'

Who Shot Rhodesian Man?

At times the contact between prehistoric primitive man and representatives of highly developed civilization appears to have resulted in a less than peaceful coexistence. While some prehistoric men were rescued from the portals of death by medicine, others not so fortunate were killed by advanced weapons.

The Museum of Natural History in London exhibits a Neanderthal skull discovered near Broken Hill, in Rhodesia, in 1921. On the left side of the skull is a hole, perfectly round. There are none of the radial cracks that would have resulted had the hole been caused by a weapon such as an arrow or a spear. Only a high-speed projectile such as a *bullet* could have made such a hole. The skull directly opposite the hole is shattered, having been blown out *from the inside*. This same feature is seen in modern victims of head wounds received from shots from a high-powered rifle. No slower projectile could have produced either the neat hole or the shattering effect. A German forensic authority from Berlin has positively stated that the cranial damage to Rhodesian man's skull could not have been caused by anything *but* a bullet. If a bullet was indeed fired at Rhodesian man, then we may have to evaluate this in the light of two possible conclusions: Either the Rhodesian remains are not as old as claimed, at most two or three centuries, and he was shot by a European colonizer or explorer; or the bones are as old as they are claimed to be, and he was shot by a hunter or warrior belonging to a very ancient yet highly advanced culture.

The second conclusion is the more plausible of the two, especially since the Rhodesian skull was found sixty feet below the surface. Only a period of several thousand years can account for a deposit of that depth. To assume that nature could have accumulated that much debris and soil over only two or three hundred years would be ridiculous.

Rhodesian man was shot by a high-velocity projectile, but the bullet that killed him must have been fired at an early period in human history.

The examination results of the Rhodesian skull are not the only evidence that someone (or even some nations) possessed rifles or similar pieces of armament in the distant past. The Palaeontological Museum of the USSR in Moscow contains an artifact that strongly supports this conclusion. It is the skull of an aurochs, a type of bison now extinct. The skull was discovered west of the Lena River, and its age has been judged to be several thousand years.

What arrested the attention of Professor Constantin Flerov, curator of the Moscow Museum, and his colleagues was that the forehead of the aurochs's skull was pierced by a small round hole. The hole has an almost polished appearance, without radial cracks, indicating that here too the projectile that penetrated the animal's skull entered at a very high velocity in a nearly level trajectory. There is no doubt that the aurochs was alive when he was shot: the calcification around the aperture is evidence of that. The distance between gunner and animal, however, was too great to inflict a mortal wound. The animal survived the wound, and died years later from other causes. But his bones lasted through the ages, and with them evidence of the destructive ability of a developed people.

CHAPTER 7

Mystery Monuments of the Builders

During the last twenty-five years, and particularly within the last decade, serious questions have continually been raised concerning the validity of the theory of evolution. They have emerged not only from such areas of research as biology, genetics, palaeontology and geology, but also from the study of archaeology, the science dealing with man's early products. All over the world, on almost every continent, stand the remains of colossal edifices of stone which, though admired for millennia, have only recently been subjected to the scrutiny of men of science in an attempt to probe the mysteries of their purpose and construction.

What has been found by these men is one gigantic mass of contradictions.

The popular view of history today is that we began in an animal existence and stumbled along over an undetermined length of time to eventually become a humanoid creature, only to pass again through successive stages of crude tool making. This last period is called the Stone Age. We are told that after all this we finally achieved civilization in Egypt and Mesopotamia, through another lengthy process of trial and error mixed with cultural invention and assimilation.

Sounds far-fetched? Yes, it does; yet this is the orthodox view of history. This view, however, is increasingly being challenged.

Rather than corroborating the concept of slow, gradual development of the arts and knowledge, a concept in line with evolutionary theory, the monuments left by our early ancestors point decidedly to an advanced technology in the remote past, which either matches or surpasses our own.

There have been, of course, many theories formulated in recent years in attempts to explain the origin of the ancient edifices, but a satisfactory explanation cannot be found until these theories are linked to those accomplishments of antediluvian technology that somehow survived into the postdiluvian era. Early man was no ape. He certainly had a well-developed knowledge of mathematics and architecture, and he must have belonged to a social order that combined the efforts for the construction of cities and for the organization of civilizations. Considering the astonishing accomplishments achieved by the first generations that survived the Flood (the Tower of Babel, world surveys, atomic power, flight, etc.), we may well wonder what advanced forms of structural and technological feats the antediluvians were responsible for prior to their being swallowed up by the waters of the Deluge.

Stonehenge Mystery Solved

Antiquity does not easily surrender its secrets, and constant probing is necessary to extract even those minute fragments of surviving knowledge which enable us to get a glimpse of our ancestors' accomplishments. But what has been discovered already only increases our eagerness to dig even deeper.

The mystery is intensified when we try to remove the obscurity from the hundreds of stone monuments that are strewn across the world, for, located on the crossings of the ley lines, these prehistoric monuments of man have been erected for a definite purpose by a race of great intelligence. Most puzzling of them all is Stonehenge, the enigmatic ring of stone standing in solitude on Salisbury Plain in southern England.

Since the seventeenth century, writers and scientists have pondered the purpose for which Stonehenge was erected, and many theories have been advanced to explain its origins. Perhaps the one man who has done more to unravel the mystery of the ring of stones is Gerald S. Hawkins, an astronomer and historian who believes that the structure is a gigantic celestial calculator. After many years of careful observation and research, Hawkins has demonstrated with the aid of a computer that the Stonehenge standing stones, or the spaces between them, were observation posts pointing to the specific points of the risings and settings of the sun, moon and stars at various times of the year. His calculations have shown that by use of the Stonehenge observatory, celestial phenomena could be accurately predicted. Stonehenge is indeed a scientific instrument of the highest order.

Diligent examination has revealed that the centre underwent three distinct waves of construction, several hundred years apart – to meet the needs of a developing society. Charcoal fragments taken from one of the chalk-filled pits, known as 'Aubrey Holes', are assigned a carbon-14 dating of 2000 BC, plus or minus 275 years. Materials removed from other holes have been dated between 2200 and 2100 BC, which suggests that Stonehenge may have been constructed almost within the first millennium after the Deluge. The second building phase, known as Stonehenge II, did not begin until several hundred years after Stonehenge was completed. Whereas the first phase had set the basic scientific philosophy for the centre, the main feature of the renewed building was the first assembly at Stonehenge of megaliths, or 'large stones'. As many as eighty-two of the five-ton bluestones were erected around the centre of the old ridge-ditch system, with the stones placed six feet apart and approximately thirty-five feet from the centre point. It appears that the stones formed a double circle, with a pattern of radiating spokes of two stones each. Since the rings were open at the north-east, facing the midsummer sunrise, the paired stones probably served as a series of observation points for the ancient astronomers.

However, it is not merely their use that constitutes the real mystery, but rather how these giant stones got to Salisbury Plain in the first place. Every archaeologist who has examined Stonehenge leaves with a different theory, but no one has been able to explain how the builders were able to transport eighty of the five-ton stones over a distance of 240 miles, crossing land and water, from Prescelly Mountain in Wales to the construction site. Nothing like this has ever been done by any other prehistoric people.

Stonehenge III heightens this enigma even more, for approximately one century aftre Phase II, between 1800 and 1700 BC, eighty-one or more stones were added to the complex, some of them weighing between forty and fifty tons. This deepens the Stonehenge riddle even more, for the source of these rocks is the Marlborough Downs, lying about twenty miles north of the complex. It has been theorized that these immense stones were moved by dragging them on wooden sledges which were rolled across log rollers. If this is what actually happened, it took from 800 to 1000 men to pull each stone, with another 200 to clear the path, guide the sledges and move the log rollers from the back of the sledge to the front. Even with efficient use of this manpower, it would have taken the builders of Stonehenge seven years to transport all the stones.

Was there perhaps another way? Is it possible that the surviving science of the antediluvians included a method of overcoming the law of gravity?

While actual proof has not surfaced as yet, there is a medieval source that may offer a clue to an alternative explanation.

The twelfth-century English historian Geoffrey of Monmouth tells, in his *Historia de Gestis Regum Britanniae*, the legend of how the great boulders of Stonehenge came to be. He reports that under the leadership of Uther Pendragon, the father of King Arthur, a force of 15,000 Britons occupied the area where the stones for the monument were to be placed. Once they had secured the land, they set themselves to the task of removing the boulders – but were unsuccessful. Even when using 'great hawsers . . . ropes . . . scaling ladders', etc., the army of men could move the gigantic stones 'never a whit the forwarder'.

Suddenly they heard a peal of hilarious laughter. Merlin the Wizard, who had accompanied the expeditionary force, came forward and, telling the men to stand aside, began 'putting together his own engines' with which he 'laid the stones down so lightly as none would believe' possible. By means of these 'special engines', the stones were transported and set up at Stonehenge, which 'proved yet once again how skill surpasseth strength'.

Geoffrey's story, of course, is a legend, but it may contain some element of truth. Simple brute force alone would have required tremendous amounts of human energy to move the stones – even if it were possible to do so at all. The stones were undoubtedly moved and transported in a special way unknown to us, and the 'engines' of Merlin may indicate that some form of prehistoric machinery provided the lift needed. The fact that modern cranes and lifting apparatus are barely able to move, let alone lift, the gigantic sarsen stones does support this.

Moving the boulders to Stonehenge was one problem; elevating them into their assigned positions may have been even more complex, for the entire observatory was built not on level ground but on a sloped surface. Measurements show that even this tilt was compensated for by the builders with an astonishing degree of accuracy.

Gerald Hawkins comments, in his book *Beyond Stonehenge*, 'Such precision of placement is, or was, astounding. To erect a boulder so that it was horizontally aligned . . . was a task difficult enough; to sink that great block into the ground just so far and no further, so that its tip was aligned vertically to an accuracy of inches, was an achievement requiring another whole dimension of skill.

'How, in fact, was it done? If, after erection, the stone had settled too deeply it would have been out of alignment – and how could it have been lifted? Of course, if it had not settled far enough its top could have been bashed away to lower it to the proper height – but the top was not bashed . . . Somehow, by a technology unknown, the Stonehengers figured out beforehand the depth of hole required to match up exactly, as far as the survey shows, with this collection of variables.'[1]

If such a task were assigned to a modern builder, Hawkins further explains, he would not be able to do so without the aid of a yard tape, plumb line, spirit levels, elevation sights, and blueprints showing the land contour and the particular design of each stone and its corresponding hole.

It is certainly apparent that the sagacious builders of Stonehenge had access to tools and instruments of precision and exactitude similar to those in use today.

The Stones and the Heavens

Since men of science down through the centuries first began to examine the boulders of Stonehenge, there have been numerous theories advanced to explain the construction's purpose; when Gerald Hawkins initiated his research he approached his story from an architectural standpoint. Touring the monument, he noted that many of the archways were very narrow, ranging from one to two feet in width. When an observer looked through two aligned archways, his view was restricted to a very small angle. It appeared that the builders had intended to limit the viewer's field of observation so that only one specific phenomenon could be seen. It seemed that the placement of the stones and archways had been made with the intention of stressing the importance of what was to be observed.

Suddenly the idea occurred to Hawkins that the viewing lines might have celestial significance. In order to test this theory, he made a meticulous record of all the possible viewing alignments through the archways.

His initial task completed, he then turned to a computer to reconstruct the way the night sky looked between the years

2000 and 1500 BC, particularly noting where certain celestial phenomena associated with the sun and moon took place. It was then just a matter of programming the computer to find whether the Stonehenge viewing alignments and the positions of the sky phenomena coincided.

The results were amazing! Twelve of the most significant Stonehenge alignments pointed, with a mean accuracy of better than a degree and a half, to important sky positions of the moon; twelve more alignments pointed to important sky positions of the sun with a mean accuracy of less than one degree. Checking further with the computer, Hawkins discovered that the probability that these Stonehenge alignments had not been planned was less than one in ten million.

There was no doubt of it: Stonehenge was built and used as a Stone Age astronomical observatory! This bizarre rock pile is actually the remains of a monumental sky computer, and with it the Stonehengers were able to predict and record with an unprecedented degree of exactness the recurring patterns of the sky and the eclipses and were also able to calculate the times and seasons for the planting and harvesting of crops.

After the erection of the fifty-ton boulders, Stonehenge was in use for roughly 500 years before it was abandoned; however, the operation of other stone computers continued, although they were smaller in size. Scattered throughout Britain are other stone rings, admittedly not as impressive in size, but equally important to the society of the builders. Stonehenge was never a unique concept – only its size was extraordinary.

In recent years Professor Alexander Thoms of Oxford University has conducted a detailed survey of over 600 British megalithic stone circles, and the dating methods employed in this study show that they were erected between 2100 and 1500 BC. Here, too, as with Stonehenge, the dates were corroborated by astronomical information.

But there were other discoveries. The study also revealed that many of the circles were laid out with a precision that today can be measured only by a highly qualified team of surveyors. For example, not far from Stonehenge, the stones of Avebury are set out with a scientific exactitude approaching 1 in 1000, while those of Penmaenmawr have an error of only 1 in 1500. This accuracy is also found on a much smaller scale, for many of the stones have cup-and-ring markings which, when carefully examined, are found to have been carved with a diameter accuracy within a few thousandths of an inch!

Primitive workmanship? Hardly! This was an extremely accomplished people, for an investigation of the 600 rings indicated that the megalithic builders laid out the various geometric forms according to an exact unit of length, what is now known as the 'megalithic yard': 2720 feet. The uniformity of this ancient unit of measurement suggests that one central authority had planned and directed the construction of all the rings.

'This unit was in use from one end of Britain to the other,' Professor Thoms concluded. 'It is not possible to detect by statistical examination any difference between the values determined from the English and Scottish circles. There must have been a headquarters from which the standard rods were sent out, but whether this was in these islands or on the Continent the present investigation cannot determine. The length of the rods in Scotland cannot have differed from that in England by more than 0·03 [inch]. If each small community had obtained the length by copying the rod of its neighbour to the south, the accumulated error would have been much greater than this.'[2]

The resulting conclusion could not be avoided. Professor Thoms says, 'The design of the necessary sectors, whether obtained by pure reason or by some complex empirical operation, demands a highly trained intellect. The discipline necessary could not have arisen out of nothing. There must have been behind it a school or system of mathematical reasoning, evidenced by the remarkable designs that we find in the complex rings.'[3]

He was simply baffled by his discovery, which was compounded by the realization that many of the ovoids, ellipses and circles were based on the use of the Pythagorean triangles, a concept which was thought to have originated with the Greeks, yet here they were, 1500 years before Pythagoras entered history.

Knowledge of the Moon 'Wobble' 4000 Years Ago

Perhaps one of the most impressive of the megalithic stone-circle sites is Callernish, situated on Lewis, the northernmost island of the Outer Hebrides, which has, among other prehistoric landmarks, an avenue marked off in stones. It is this stone avenue that has currently become the focal point of a new discovery. As seen from Callernish, the midsummer moonset occurs over Mount Clisham, and the avenue points directly towards the mountain. Because the Callernish complex lies only 1·3 degrees south of the arctic latitude for the moon, the megalithic observers would have seen a peculiar phenomenon: once every eighteen or nineteen years the moon would appear to stand still about one degree above the horizon. This eighteen/nineteen-year cycle is, of course, the same as that recorded at Stonehenge. The avenue stones are aligned in such a way that the prehistoric astronomers were able to observe what is called the moon's wobble – the small amplitude ripple of the moon's declination at extreme positions. Before Callernish was investigated, it was believed that this phenomenon was not discovered until the sixteenth century, by Tycho Brahe. The period of the wobble is 173 days, and the wobble reaches its maximum amplitude immediately before the season for lunar eclipses! The Callernish builders, it appears now, possessed a unique computer in stone for predicting lunar eclipses.

Another significant point to note is that many of the alignments of Callernish are the same as those found at Stonehenge, with the key observation stones laid out in a very similar geometric pattern. Callernish is situated at a latitude where the moon appears to skim the horizon; Stonehenge is also located at a spot where the extreme positions of the moon appear at right angles to those of the sun.

If Callernish and Stonehenge are related works – and the fact that they used the same basic measuring unit in their structures would tend to confirm this – then the builders were aware of the differences in the celestial phenomena observed at both structures – differences which could easily have led to a knowledge of the curvature and size of the earth.

Other Megalithic Sites in Britain

Even though the builders were engaged in massive construction projects enabling them to chart the course of heavenly bodies, this was hardly their only endeavour. At the same time that Stonehenge and Callernish were being built, other remarkable projects were undertaken. One of the most common was the long barrow, or burial mound. Even though the greatest concentration (350) of these is found in the Salisbury area of England, the most outstanding one is located at West Kennet, about sixteen miles north of Stonehenge. Constructed long before 2000 BC, this mound is 350 feet long and varies in width from 75 feet on the east end to 50 feet on the west, where it terminates in a sepulchre approximately 35 feet wide, 43 feet long and 8 feet high. The entrance was blocked by several enormous stones, one of which weighed about 20 tons. What is most fascinating about this West Kennet barrow is that when dating techniques were applied to it, it was found to be one of the oldest, if not the oldest barrow in Britain. Yet it demonstrates building skills of the highest order.

Excavation of the barrows has provided many surprises. They have shattered the belief that the earliest Britons were isolated from the rest of the world, because in fact their communication with the Continent and the Mediterranean area was much greater than that of the Britons of several centuries later. Among the remains in the tombs were bronze pins from Bohemia, faience beads from Egypt, and amber from the Baltic.

The builders surpassed the stone circles or burial mounds, for half a mile beyond West Kennet is the largest artificial earth mound in Europe, Silbury Hill. The reason for its existence is still a mystery, although scientists now claim to be inching closer to the truth. Conical in shape, it rises to a height of 130 feet, with a circular base more than 200 yards in diameter. It covers 5·5 acres and its total volume has been estimated at 405,000 cubic feet, and its construction may have required as much as 2 million man-hours – thought to be even greater than that needed for building Stonehenge.

Many explanations have been given for the erection of this massive structure, the first being that it was a huge burial mound; however, excavations into the top and sides have revealed no funerary or skeletal remains. Today the most accepted theory is that the great hill, like Stonehenge, was designed to measure celestial phenomena, for there are indications that a large maypole may have topped the hill and that the shadow the pole cast was used to calculate the length of the year. Invariably the monuments erected in that period point to definite connections with celestial observations, yet there is at least one known exception. This monument, noted not for its great height but for its length, is one of the greatest engineering feats accomplished by the British megalithic builders. From Salisbury Plain, beginning at the southern end of the Avebury stone circle and extending for two hundred miles north-eastward to Norfolk, is an extraordinary prehistoric highway called the Icknield Way. The road runs dead straight on level ground and follows perfectly the contour of the land in hilly areas. It is level and widens out in some places to the equivalent of a modern four-lane highway. It is superior to any road constructed by the Romans, yet it predates the Romans by 2000 years. Why did the megalithic builders need a highway, when archaeologists believe that they did not even have the wheel?

Europe, Africa and the Middle East Not Excluded

Communications must have been good between England and the Continent, and the roadways and sea lanes were undoubtedly well travelled, for the remains of the megalithic monuments were not limited just to the British Isles. They are found scattered across the globe. Stonehenge may have been the focal point of activity, but from there the builders, architects and astronomers fanned out over the entire world, leaving monuments wherever they went. Across the Channel from England, in the French province of Brittany, there are several megalithic sites. They are also found at Kerlescan and Kermario; in fact, within a distance of 3250 yards, there are nearly 3000 menhirs (single standing stones), most of them in rows pointing towards long-forgotten grave sites and facing the midsummer sunrise. The remains of the chieftains have long since become part of the earth, but their tombs endure, a testament to their greatness.

Elsewhere in Brittany there are other megalithic monuments, some of which are constructed from the largest standing stones on record in western Europe. The menhir of Ile-Melon, unfortunately destroyed during World War II, originally weighed ninety tons. The largest was the 'Fairy Stone' of Locmariaquer. Broken up by lightning in the eighteenth century, it once stood sixty-seven feet high and weighed over 380 tons!

But, again, Britain and France are not the only countries where the builders left their marks. Far beyond Brittany, on the coasts of Germany, Holland, Scandinavia, Portugal, Spain, the Balearic Islands, Corsica, Sardinia, Sicily and Malta, at Tiryns and other Mycenaean sites, there is ample evidence of the past work of the ancients. The grave sites and the stone circles all testify to the skill of the builders. Their tracks are found even in North Africa and the Middle East, telling us of their far-reaching wanderings and of the spread of their civilization. In Morocco, dolmens (a circle of stones capped by a larger stone) are found in the district of Kabylia; a stone circle is found near Tangiers. Other dolmens have been discovered in Algeria, while Libya, Syria, Jordan and Lebanon have literally hundreds of circles and free-standing stones, all testifying to the builders' presence at one time.

And then there's Egypt. Stretched along the Nile, the sandy

countryside of the land of Amen-Ra is speckled with the remains of dolmens which mark the sites where the ancient people buried their dead and which were subsequently joined by the tombs of the pharaohs.

Three sites in the Middle East are of particular interest because of advanced scientific and engineering skills involved in their construction. At Baalbek, in modern Lebanon, the Romans constructed their magnificent temple to the sun, a temple which was dwarfed in size, however, by the immense prehistoric dressed-stone platform on which it was built. Of unknown age and origin, the platform is a feat of engineering that has never been equalled in history. It is made of individual stones eighty-two feet long and fifteen feet thick which are estimated to weigh between 1200 and 1500 tons each. Of the stones cut for the platform, the largest one was not transported to the site but instead was left at the quarry half a mile away. Called Hadjar el Gouble, or 'The Stone of the South', it weighs more than 2000 tons. There are no cranes or other lifting apparatus in the world today that can budge, let alone lift, the titanic blocks of Baalbek – yet there they are, cut and fitted together with such precision that a knife blade cannot be inserted between the blocks.

The second site, equally remarkable, is located on the wind-swept moor of the Golan Heights in Israeli-occupied Syria. There Israeli archaeologist A. Itzhaki recently uncovered the remains of five giant stone rings believed to be a thousand years older than Stonehenge. A line drawn through the area where the rings overlap points to true north. Because of the unreliability of compass readings in the vicinity of basaltic rocks, the engineering skills required to find true north were of a degree of skill generally considered beyond the reach of the ancients.

The third site is far to the north, at Medzamor in Soviet Armenia, where the Russian scientist Dr K. Megurtchian has discovered what is thought to be the oldest large-scale metallurgical factory in the world. In close proximity to this, geometric patterns that were found cut into the volcanic rock point to various celestial phenomena. One distinct line points to the spot on the horizon where the star Sirius rose between 2600 and 2500 BC.

What is especially intriguing about the Medzamor site is that it is located only fifteen miles from Mount Ararat, the historical and legendary landing place of the only survivors of the ante-diluvian civilization.

Did the Megalithic Builders Reach the Americas?

With the passing of time, the controversy over who really was the first to discover America becomes more intense, as if it actually mattered. For years there have been pitched verbal battles among renowned historians, lengthy intellectual discussions, and countless magazine articles, all hoping to solve this riddle. Was it Columbus? Could it have been Leif Erikson? Still other names have been proposed and just as rapidly discarded again. Perhaps the answer lies somewhere else – on a prehistoric site called Mystery Hill in North Salem, New Hampshire, where twenty-two large stones stand majestically on top of a 200-foot-high hill. The origin and significance of the site are shrouded in darkness; its age, however, is not. Carbon-14 tests conducted in 1969 date Mystery Hill between 1225 and 800 BC, long before the arrival of the Indian tribes that once inhabited the area, but in the same time slot as similar megalithic constructions in southern Europe. Mystery Hill suffered partial destruction during the eighteenth and nineteenth centuries, when some of the stonework was removed and used to build a nearby sewer system.

The stones of Mystery Hill are arranged in an elaborate system of tunnels, menhirs and dolmens, and have just recently been found to be celestially aligned. Each year on the first day of winter, for example, the sun, when viewed from the centre of the hill, sets directly over what is called the Winter Monolith.

Were the builders the first to leave their prints on the sandy shores of the eastern coast of America? History remains silent when confronted with this question, but the hill was definitely built by someone, and its similarity to the megalithic sites of Europe is more than coincidental.

Mystery Hill is not unique, however, for other sites not unlike this prehistoric perplexity exist in different parts of the Americas. On the Central American island of Bonacca, archaeologist F. A. Mitchel-Hedges discovered an ancient 800-yard-long wall enclosure with two large standing stones reminiscent of those found at Stonehenge. The stones measured approximately 7 feet in height by 2½ feet in diameter. Also discovered were a number of oddly shaped carved stones that appeared to be older than the Mayan, Toltec and Aztec civilizations. An even more startling find was made at La Venta, at Villahermosa, Mexico, where there are menhirs and troughs in long alignments that strikingly resemble the rows of stones found in Brittany. Near the prehistoric fortress of Sacsahuaman, Peru, on a rocky spur called K'emko, menhirs and other roughly hewn stones have been found, once again corresponding in appearance to the European monuments!

Orthodox historians unfortunately have done very little to take note of the important megalithic sites found in the New World, as their acceptance would disturb long-cherished theories. They simply cannot account for the fact that a prehistoric race such as megalithic man could have crossed the Atlantic and left its mark in America, when supposedly more civilized later people were unable to do so.

It is with the same closed-mindedness that the historians look at the discoveries made in Asia and the Pacific, where remains of the builders' activities have surfaced in the most unexpected regions. In India dolmens dot the land from the Nerbuddha River to Cape Comorin. At latest count, the Neermul jungle of Central India has yielded at least 2000 of the monuments it has hidden for centuries, and another 2200 have been located in Dacca.

Monuments of a similar nature have also been found in China, Korea and even Japan. The mystery of the builders' activities increases as the geographical boundaries expand. On the south-eastern shore of Ponape, in the Senyavin Islands of Micronesia, the remains of a huge temple complex called Metalamin face the midsummer sunrise. There is every indication that in the days of the builders, the population of Ponape Island was many times what it is today, for Metalamin is sufficiently large to seat as many as two million people! The ruins, like those in Europe and America, are composed of vast stone blocks weighing as much as fifteen tons each. These blocks were transported from a quarry approximately twenty miles away – with not a hint of how this was accomplished.

Were they navigators as well as builders? History stands mute on this question, but the fact remains that three thousand miles away, south-east of Ponape, on tiny Malden in the Line Islands, is a second group of ruins architecturally similar to Metalamin! There is, however, one important difference. The ruins on Malden are connected to the rugged coastline by a number of prehistoric basalt-paved highways, a situation which baffles the scientists.

'But they can't be highways,' the archaeologists cry out in despair. 'These people didn't have the wheel . . .'

Oh, didn't they? There are still many ruins the builders have

171

left on other Pacific islands, but most of these are still being excavated. One can say without hesitation, however, that the most famous and mystifying of all the Pacific monuments are those strange statues that stand in peaceful silence on a lonely rock called Easter Island.

Unresolved Mystery of the Stone Faces

There are few detective stories as confounding as the one that came to the attention of the western world on Easter morning 1722, when the Dutch explorer Jan Roggeveen first set eyes on a tiny speck of land in the broad expanse of the Pacific Ocean. Unable to locate it on the navigational charts, he christened the new-found territory Easter Island. With the anticipation of finding treasure, he anchored his ship and rowed the few hundred yards to the rocky shore. But he soon realized that the volcanic isle had little to offer. There were no trees and no indigenous animals, and Roggeveen found only a few hundred scantily clad natives dwelling in huts along the jungle-fringed beaches. The island was barren and inhospitable, yet it did give one thing to the world: a mystery unrivalled anywhere in the vast Pacific. Scattered over the rocky ground, strewn about the meadows of sparse grass and sullenly peering from the slopes of the island's volcanoes were hundreds of stone faces jutting out of the soil, each with the same mute and meaningless expression, long straight nose, narrow and tightly closed lips, sunken eye sockets, and low forehead.

Who made them? Where did they come from? What was their significance? Roggeveen and his crew undoubtedly gaped at them in utter bewilderment, for nothing like this had ever been encountered. The statues were certainly not the kind he would carry back to Amsterdam as trophies of a discovery voyage. There was something weird, something eerie about them, and the sombre expression on their stone faces became an ever-returning topic of conversation on the long voyage home.

More than 250 years have passed since that day, and Jan Roggeveen is now merely a name written on the pages of history books, but the secret of the silent statues still continues to evade us.

The enigmatic question surrounding the Easter Island statues is not what they are supposed to represent, but rather how they were moved from their quarry at the edge of the volcano Rano-Raraku to their present sites, a distance of up to five miles. In

1956 the Norwegian explorer Thor Heyerdahl, known for his Kon Tiki expedition, visited Easter Island to conduct the first large-scale investigation of the statues and their history. He soon realized that discovering their origin was not half as challenging as solving the problem of how the monstrous heads had been transported and erected. Convinced that the builders had nothing but brute manpower at their disposal, he contracted a dozen island natives to employ muscle to move a stone head. With steadily increasing frustration, the team laboured for eighteen days, using the 'heave-ho' method, which at last enabled them to set up one of the heads. This answered the question for Heyerdahl; satisfied that he had found the solution, he abandoned the project. His efforts are now being cited in many scientific journals, but did he really duplicate the way it was done?

There are several objections to Heyerdahl's experiment which cannot be ignored. What is generally not known is that the statue chosen for the project by Heyerdahl's men was not the average Easter Island statue, for the weight of the island heads is roughly between 35 and 50 tons each, but the head which was arduously moved by twelve sweating natives weighed somewhere between 10 and 15 tons. Granted, it was still a momentous achievement, but the result did not qualify it as a 'typical' example. Second, Heyerdahl's stone was transported only a few hundred feet, across smooth sandy ground that exists only at Anakena, the place from where the statue was moved. The contrast between Anakena's terrain and that of the rest of the island is too great, because the area over which the other stone heads had to be transported consists of volcanic rock, which is hard and uneven. If the heads had indeed been dragged across this surface by Thor Heyerdahl's proposed method, then the stone statues would have been grooved with long scars. None of the statues reveal any such markings.

The type of equipment used in the moving process presents another problem. Heyerdahl's natives utilized ropes and wooden poles to aid them in erecting and manoeuvering their statue – but there originally was no wood on Easter Island. Currently sycamore trees grow on the island, but only because some of the early European settlers brought them there. The records of Jan Roggeveen do not mention trees, and Captain Cook also noted an absence of trees upon his arrival on the island. If wood was indeed used by the builders, then they must have imported it by ship from the nearest forest – 2500 miles away. As for the ropes, Heyerdahl's experimental team used sturdy, well-manufactured ropes from Europe. It was fortunate for them that they did, for ropes made from the indigenous reeds of Easter Island were neither strong nor durable and were most certainly not adequate to the job. Heyerdahl's moving of one single stone head over a flat and relatively even surface also had no bearing on how other heads were moved up and down cliff walls, as there are many spots on the island where this did occur in the days of the builders. At the quarry of Rano-Raraku, twenty-ton statues were carved near the top of the crater, then lowered 300 feet, over the heads of other statues. This was accomplished without leaving even a mark. The stone heads sitting on ledges in the cliff of Ahu-Ririki are the best illustration of this operation. Here the sheer rock face plunges 1000 feet, straight to the sea. The gusty winds at the top are usually strong enough to blow a man off balance, while the sea currents below are so treacherous that a boat cannot approach the rock. Yet at an elevation of 600 feet on the cliff wall stands a platform that bears the marks of a number of twenty-five-ton statues, the remains of which now lie on the ocean floor. Heyerdahl may have moved one small head; he has yet to present an answer that can withstand scientific test.

But he was not the first to fail. In the late nineteenth century, the French ship *La Flore* visited Easter Island with the intention of taking one of its statues back to Paris. It took a 500-man work force to carry the 7½-foot-tall statue, one of the smallest of the island heads. Today, much battered and bruised by its ordeal, it can be viewed at the Musée de l'Homme.

Even though these two heads have been moved, the question still lingers: How did the builders of Easter Island cut, move and erect the gigantic heads, including those which approach the size of a seven-storey building?

There have been many theories about who actually created and erected the solemn stone heads of Easter Island, but there are no easy answers. Their origin has been attributed to nearly everyone, from survivors of the so-called lost continents of Mu and Lemuria, to tribes of wandering Polynesians who supposedly sculptured the monolithic monstrosities to while away their idle hours.

What, then, is the answer? Evidence of a more realistic possibility is found in a group of stone buildings which few modern explorers and researchers have diligently investigated. Thirty-nine in number, they are located on Easter Island in Orongo. Each structure is oval in shape, measures approximately seven yards in length and two yards in width, and is topped by a low circular ceiling. The foundation stones were laid beneath the surface and were followed by rings of stone blocks, with each ring narrowing towards the centre until the sides converged in the rounded roof. Francis Maziere, one of the few western experts who have visited and described the ruins, was impressed with only one point: that these stone buildings are nearly identical in shape and construction with those erected by the builders in the Mediterranean area! For those who still wonder about this connection, there is one more feature linking these ruins with those of the European structures. At the Orongo site lie the remains of a small solar observatory, composed of one or possibly more standing stones, by means of which the ancient observers were able to calculate the movements of the sun. Was this perhaps the beginning of a Stonehenge which was later abandoned?

There is no decisive evidence that the men of prehistory who erected Stonehenge were also responsible for creating the stone heads of Easter Island, but, judging from the ruins, it is obvious that the two sites are parallel, not only because both were constructed from stone blocks but also because their building techniques were similarly advanced.

What Happened at Tiahuanaco?

Two thousand miles north-east of Easter Island, high in the Andes mountains of Peru, on the picturesque shores of Lake Titicaca, stand the remains of a city of startling dimensions – and no one knows its origin. Not even the oldest living Indian could tell of its history when questioned by the Spanish conquistadors in their bloody assault on the area in 1549. Whoever its engineers were, they certainly were not related to the Indians in any way, as the foreign element is quite obvious both from the style of the structures and from the fact that the statues of Tiahuanaco depict strange-looking men with beards, not the usual Indian faces which tend to be devoid of beard growth.

The society that developed the entire Tiahuanaco area had technical abilities that astounded the conquistadors. Archaeologists who have studied the site since its discovery by the Spaniards have uncovered features thought to be unknown to the ancients. The Akapana, or 'Hill of Sacrifices', one of the three important temple sites, was a huge truncated pyramid, 167 feet high, with a base 496 by 650 feet. The now-crumbling sides of the impressive structure were perfectly squared with the cardinal points of the compass, a feature common with other great edifices found around the world, including the Great Pyramid of Gizeh. The destructive plundering of the Spanish conquerors erased clues which might have served as keys to unlock the secrets of the ancient inhabitants, and the ravages of time have done the rest. Today the side surfaces of the Akapana are rough and torn; the stone slabs that provided a protective cover for the mound have been hauled away and used in construction projects. An enormous stone stairway that once flanked the hill has also become a victim of gross vandalism. Today, only a few steps remain. The reservoir system that once topped the Akapana indicates the high degree of development of the builders. The hill still reveals evidence of the precision-designed, intricately cut stone conduits and overflow pipes, especially graded to ensure the proper flow of water Similar pipes are found scattered throughout the Tiahuanaco complex, suggesting that the city had a complete drainage, water supply, or sewage system.

But other probes have extracted still more from the Andes. A thousand feet north of Akapana is the Curicancha, or 'The Temple of the Sun'. It rests on a stone platform 10 feet high and 440 by 390 feet on a side, composed of blocks weighing 100 to 200 tons each! The walls of the temple complex itself are constructed of blocks weighing 60 tons each, while the steps of the stone stairway weigh an impressive 50 tons apiece. Other structural units, composed of 200-ton blocks, lie haphazardly, just where they fell. Tiahuanaco is a place where contradictions and impossibilities reign supreme. Things that can't happen have happened here. It's amazing that the city exists at all: the entire metropolis was built 13,000 feet above sea level, and the air pressure at that altitude is only eight pounds per square inch, as compared to fifteen pounds at sea level. The thin, oxygen-poor air sears the throat and nose, and even the slightest exertion may cause nausea, headaches, and sometimes even heart attacks. In addition, no seeds will sprout or grow at that elevation, which means that there was no local food supply to support a large working crew. Yet somehow, under extremely hostile conditions that threatened life itself, the builders managed to manoeuvre hundreds of stone slabs weighing up to 200 tons each into their predetermined places. The quarry sites of the stones have been discovered on an island in Lake Titicaca, but near the shore opposite Curicancha. It was therefore necessary to transport the stone over distances ranging from thirty to ninety miles. In rarefied air the movement of massive objects over such great distances is not possible by muscular strength, but the stones were moved nevertheless and found resting places in Tiahuanaco.

If muscular energy was not sufficient, then what was used?

The Mystery Fortresses of the Andes

Tiahuanaco is by no means unique, for scattered throughout the Andes are several fortresses of very similar design, all predating the ancient Incas by an unknown period of time.

In Chile, high on the plateau of El Enladrillado, 233 stone blocks are placed geometrically in an amphitheatre-like arrangement. The blocks are roughly rectangular, some as large as 12 to 16 feet high, 20 to 30 feet long, and weighing several hundred tons. As at Tiahuanaco, huge chairs of stone have also been found in disarray among the ruins, each weighing a massive 10 tons. Perhaps the most important find at El Enladrillado was the discovery of three standing stones at the very centre of the plateau. Each is 3 to 4 feet in diameter. Measurements reveal that two of the stones are perfectly aligned with magnetic north, while a line through one of these and the third stone points to the midsummer sunrise. Were the builders here, too?

To the north, at Ollantaitambo, Peru, is another pre-Inca fortress, with rock walls of tightly fitted blocks weighing between 150 and 250 tons each. Most of the blocks consist of very hard andesite, the quarries for which are situated on a mountaintop seven miles away. Somehow, at an altitude of 10,000 feet, the unknown builders of Ollantaitambo carved and dressed the stone (using tools, the nature of which we can only guess, that could penetrate such hard rock), lowered the 200-ton blocks down the mountainside, crossed a river canyon with 1000-foot sheer rock walls, then raised the blocks up another mountainside and placed them in the fortress complex. As South American antiquarian Hyatt Verrill notes, no number of men – Indian or otherwise – could duplicate this feat with only stone implements or crude metal tools, ropes, rollers and muscle power. 'It is not a question of skill, patience and time,' Verrill explains. 'It is a human impossibility.'

Is it possible instead that a higher form of prehistoric technology was employed, of which we know absolutely nothing?

One of the most impressive 'mystery fortresses' of the Andes is Sacsahuaman, located In the outskirts of the ancient Inca capital of Cuzco. It rests on an artificially levelled mountaintop at an altitude of 12,000 feet, and consists of three outer lines of gargantuan walls, 1500 feet long and 54 feet wide, surrounding a paved area containing a circular stone structure believed to be a solar calendar. The ruins also include a 50,000-gallon water reservoir, storage cisterns, ramps, citadels and underground chambers.

What is truly remarkable about Sacsahuaman is the stonework. Here extremely skilled stonemasons fit blocks weighing from 50 to 300 tons into intricate patterns. A block in one of the outer walls, for example, has faces cut to fit perfectly with twelve other blocks. Other blocks were cut with as many as ten, twelve, and even thirty-six sides. Yet all the blocks were fitted together so precisely that a mechanic's thickness gauge could not be inserted between them. And even more extraordinary is the fact that the entire Sacsahuaman complex was built without cement.

As with the other mystery fortresses, the question of how the stones of Sacsahuaman were transported remains unanswered. The quarries from which the stones for Sacsahuaman were brought are located twenty miles away, on the other side of a mountain range and a deep river gorge. How the massive stones were moved across such hopeless terrain is anyone's guess.

Sacsahuaman poses many mysteries, yet it possesses one more which few orthodox historians are willing to recognize or study because of its 'impossibility'. Within a few hundred yards of the Sacsahuaman complex is a single stone block that was carved from the mountainside and moved some distance before it was abandoned. An earthquake apparently interrupted the progress of the movers, for the stone was turned upside down and is damaged in several places. It contains steps, platforms, holes and other depressions – a masterpiece of precision cutting and dressing, clearly intended to become a part of the fortification. What is truly impossible about the block is that it is the size of a five-storey house and weighs an estimated 20,000 tons! We have no combination of machinery today that could dislodge such a weight, let alone move it any distance. The fact that the builders of Sacsahuaman could and did move this block shows their mastery of a technology which we as yet have not attained.

The Lines of Nazca Valley

The Andes conceal many ancient wonders of construction, not all made of stones hauled across inconceivable distances. Not far from the Pacific Ocean, in the Peruvian foothills of the Andes

178

250 miles south of Lima, is the historical city of Nazca. It is of important archaeological value; however, the city's real curiosity is not its relics but the valley in which it lies – a strip of level desert ground thirty-seven miles long and a mile wide. Covering nearly every acre of the Nazca Valley are enormous drawings scraped out on the desert floor – lines running in all directions: elongated cleared areas, spirals, zigzags, birds, spiders, monkeys, snakes, fish, etc. They were revealed by removing the dark purple granite pebbles which lay on the Nazca desert and exposing the light yellow sand just below the surface. Since there is little rain or wind erosion at Nazca, the lines and figures have remained intact for an undetermined number of centuries. Yet during most of that time, travellers trekking through the valley never noticed the drawings, because unless one is standing directly on one of the lines, the areas where the pebbles have been scraped away are not noticeable. Move a few feet away, and the line blends into the rest of the rough desert terrain.

Not until the 1930s, when the first commercial airlines began operating over the Andes, did sightings from the air confirm the existence of the Nazca drawings. Obscure on the ground, they are clearly seen from above – clearly enough, in fact to have been viewed by the astronauts aboard Skylab, orbiting 270 miles above the earth. Yet there is no high mountain, plateau or other natural elevated point nearby from which the Nazca artists themselves could have seen the drawings in their true perspective. So why were they made? Did they serve some purpose? Did the artists also perhaps master the art of flight?

The first detailed study of the Nazca mystery was initiated in 1946 by the German astronomer and archaeologist Dr Marie Reiche, who devoted the next twenty years to taking accurate surveys of the ancient drawings and speculating on their significance. Dr Reiche focused her attention at first on the numerous lines criss-crossing the valley. Many of these, she discovered, ran straight and true for up to five miles. Some are parallel to one another; others gradually converge, while still others radiate from specific points – small mounds of boulders. Dr Reiche even discovered lines which appeared to run straight into the bases of mountains and emerge on the other side in complete alignment and at the same level. When the degree of straightness of the Nazca lines was checked by modern measuring equipment, a startling observation was made: the average error was no more than nine minutes of arc, a deviation of $4\frac{1}{2}$ yards per mile. That figure is the limit of accuracy that can be obtained by what is called photogrammetric survey. In other words, the ancient lines were laid out straighter than can be measured by the best of modern survey techniques. Dr Reiche stated, 'The designers, who could only have recognized the perfection of their own creations from the air, must have previously planned and drawn them on a smaller scale. How they were then able to put each line in its right place and alignment accurately over large distances is a puzzle that will take us many years to solve.'

It is the opinion of Dr Reiche and several other students of the Nazca enigma that some of the lines may be aligned with the risings and settings of the sun, moon, and possibly several bright stars. In fact, recent investigations showed that thirty-nine lines do point to solar or lunar events and that seventeen are associated with the stars. But this is only a small number; the majority of the lines have no celestial significance, and their purpose remains a mystery.

The Nazca Artists – Their Knowledge of the World

As extraordinary as the lines are, the details of the many animal figures etched out on the Nazca Valley floor are equally as remarkable. One of the most puzzling is the picture of a spider, 150 feet long, drawn with a single continuous line half a mile in length. What is so peculiar about the spider is that one of its legs is deliberately lengthened and extended, and at the tip there is a small cleared area. There is only one spider known that uses the tip of its third leg in precisely the manner depicted in the desert drawing, and that is the *Ricinulei*, which lives in caves deep in the Amazon jungle, a thousand miles from Nazca. Known to scientists for its unique method of copulation, for which the spider uses that extended leg in the described manner, the *Ricinulei* is extremely rare. Its mode of reproduction can be observed only with the aid of a microscope.

How the Nazca artists were able to find and then observe their tiny model we cannot say, unless we ascribe to them a knowledge of science equalling our own.

There are several indications, both from the valley etchings and from remains of Nazca pottery found in the immediate desert area, that the ancient artists had knowledge of the world far beyond the horizons of Nazca. One desert drawing depicts a thin-limbed monkey, recently identified as the spider monkey, another inhabitant of the distant Amazon jungle. On one remnant of a Nazca pot is a distinct picture of a white-breasted, black-coated penguin. The difficulty here is that penguins are indigenous to Antarctica – nearly 6000 miles away, although they are living in the Galapagos Islands. How could the Nazcans have drawn the birds unless they had actually seen them?

The most startling picture of all, however, was found on another piece of Nazca pottery, which showed faces of five girls – one white, one red, one black, one brown and one yellow. These colours could not have been chosen fortuitously, as all the races of man have been clearly represented. The faces seem to indicate that the Nazcans had knowledge, possibly even models to work from, of each and every racial group around the world. Could this be evidence of global communication in the distant past that equalled that of modern times?

As the study of Nazca progresses, more questions have arisen than can be answered. When were the Nazca drawings made? A wooden post was discovered at the intersection of two of the Nazca lines, and carbon-14 tests revealed a date of AD 500. From this, orthodox historians have ascribed a relatively recent date to the Nazca drawings: between AD 200 and 700. But it is not known whether the post was placed while the lines were being made or after they were finished. There is no way, in fact, to date the lines themselves, and it is entirely possible that they could be thousands of years older. How were they constructed? The accuracy of the drawings over such a large area attests to a remarkable engineering skill not previously believed possible for any ancient people. There is a question not only of advanced knowledge, but also of performance: the planning, engineering and construction of the drawings would have required the energies of a large number of workers. There is no water, food or shelter anywhere in the desert valley of Nazca that could have provided the necessities of life for so great an undertaking. So how was it accomplished? And the most perplexing question of all, why? Why were the drawings made in the first place? For that we have as yet no satisfactory answer.

The Great Pyramid – The Great Enigma

It is not possible to discuss the profound knowledge of the ancients without letting the mind drift in the direction of the land of Amen-Ra. I recall endless lectures in Egyptology and animated discussions on the role of the gods in Egyptian history. I also remember long winter nights in the Egyptology room of the university when I fought my way through Sir Alan Gardiner's

Egyptian grammar, deciphering funeral texts on ornate caskets stolen from the graves of the pharaohs and their nobles. But nothing really prepared me for the wonder and awe I felt when I first viewed the pyramids from atop a swaying camel.

Coming face to face with the witness of history known as the Pyramid of Cheops is an incomparable experience. Standing on a rocky, artificially levelled plateau about ten miles west of modern Cairo and not far from the rotting circus tent that houses the Gizera nightclub, the Great Pyramid has silently beheld many battles fought within its shadow during its 5000-year history. But perhaps the greatest battle of them all is the controversy raging between orthodox historians on the one hand and archaeologists, statisticians and more liberal-minded historians on the other, over the questions posed by the pharaoh's tomb, for with each new year added to its history, the slumbering giant becomes more puzzling.

The questions confronting science in connection with the tomb of Cheops are multiple and are all related to the construction of the 2,300,000 blocks weighing an average of $2\frac{1}{2}$ tons, with the largest of them – found in the roof the King's Chamber, a dark musty-smelling room in the heart of the structure – weighing over 70 tons each. Comparison of the blocks with the quarries in Egypt has confirmed the theory that the stones were brought to the site from a few miles away at Mokattan as well as from 500 miles south at Aswan.

Here, too, we face a problem when following in the tracks of the builders. How were the blocks transported to the building site and, almost equally important, how many workmen were required to move them and how long did it take?

Guesswork will not suffice in ascertaining the truth about these crucial points, for these problems are real.

I recall from my early studies that orthodox historians spouted forth the same set of answers: Quarry inscriptions on a number of the blocks ascribe the building of the pyramid to the Pharaoh Cheops in the Third Dynasty of the Old Kingdom. Since his reign lasted only twenty-two years, this would suggest a maximum time period during which the structure was erected. The blocks were either transported on wooden sledges or floated down the Nile on wooden rafts. It is further believed that 100,000 men, working for twenty years, completed the task of building the pyramid.

Fantastic? Not to the historians, for this is what is believed and what is currently taught. After all, how can one expect great efficiency from a nation whose citizenry was only one step beyond the cave-man stage? As credible as it may seem to the historians, this simple solution will certainly not resolve any of the outstanding questions. The historians are concerned only with history, not with logistics; yet that is where the answer lies.

Let's look at a few basic statistics. If 2,300,000 blocks were placed in the pyramid in 20 years' time, that is, in 7300 days, then we must assume that an incredible 315 blocks were positioned each day, or 26 blocks per hour per 12-hour day. With 100,000 men, utilizing the most modern construction equipment available today, our engineers would not be able to match this 'primitive' accomplishment. In addition, since nine months of the year were customarily set aside for planting, cultivating and harvesting, the work force could have spent only three months out of every year on the construction site. Thus, even at the exceptional rate of 315 stones per day, the amount of time spent in building the pyramid would have been eighty years, not twenty.

The famed Egyptologist Sir Flinders Petrie has estimated that eight men might have been able to handle 10 of the $2\frac{1}{2}$-ton blocks in the required three months. Using only ropes and wooden levers, it would have taken them six weeks to pull the stones out of the quarry, another week to float them down the Nile, and still another six weeks to drag them to the base of the pyramid. Eight men moving ten blocks means that 100,000 men could have transported 125,000 blocks a year, completing the massive construction project in the proposed twenty years. But this increases the number of blocks to 1500 per day – an impossiblity even by modern standards!

Manpower is another area that presents a problem. The 100,000 man labour force mentioned above is only the estimated size of the transportation crew. Add to this another 100,000 stonemasons at the quarries; 100,000 builders at the pyramid itself; still another 100,000 architects, planners, and supervisors co-ordinating the project; 250,000 women and children preparing meals and keeping shelters in good repair; and a standing guard force of 300,000 policing the workers and keeping order among them, and we are speaking of a project that required almost one million people – in the total construction – one-third to one-half the estimated population of all of Egypt around 2700 BC.

Does this sound even remotely reasonable? Not really; yet this is what we are being taught at the universities of the world. But to continually call upon the energies of a million people, year after year for twenty years, is stretching credibility to the limit.

. Some maintain that the workers were mere slaves and did not really detract from the native Egyptian labour force, but here too we run into a snag. Herodotus, who visited Egypt in ancient times and recorded its history, tells us that the Egyptians were paid for their services in building the pyramids in wheat, beer and other foodstuffs. What ruler could have paid one million workers for three months' labour every year for twenty years without going bankrupt? And where would he have obtained the immeasurable quantity of food with which to pay them?

The source from which we gather much of our knowledge about Egyptian history has been the hieroglyphic inscriptions and tomb paintings. Many orthodox historians use these tomb paintings to support their improbable claim that the building blocks for the pyramids were either hauled or floated, or both. To substantiate their claims, they direct us to two tomb paintings, one in the Twelfth-Dynasty tomb of the nobleman Djehutihotep, the other in the funerary santuary of Queen Hatshepsut. The first shows a statue being drawn on a wooden sledge pulled by 172 men, over ground which has been purposely dampened. The second picture depicts a number of Queen Hatshepsut's royal barges, which were used to float stone obelisks down the Nile. Each barge, it appears, had a displacement of about 1500 tons.

On the surface this seems to provide adequate material to defend the historians' position, but a closer examination of the facts completely repudiates this. The objection is that the two tomb paintings were made a thousand years after the pyramid was built. Sledges and barges may have been used to transport heavy objects in the Twelfth Dynasty and later, but we are concerned with methods employed in the Third Dynasty, not in the Twelfth. There is no concrete evidence that these methods were used in the construction of the Great Pyramid. In addition, we are referring not merely to the transportation of a few heavy statues, but to the logistical problem of moving 2,300,000 blocks. If for argument's sake we want to believe in wooden sledges and barges, from where would the voluminous supply of wood come? The trees of the Nile Valley are date palms, a vital source of food that could not have been spared. The wood therefore must have been imported. We know from the Egyptians' records that as early as 2800 BC they were importing large quantities of lumber from the Lebanon, the ancient world's major source of cedar wood. Considering the need and the size of the average Lebanese cedar, mathematicians tell us twenty-six million trees would have been required to fashion the necessary number of sledges and rafts. Neither the Lebanon nor all the forests in the ancient world could have supplied that much wood in twenty years, whether or not there was a fleet that could carry it all!

The truth is that it did not take twenty years to build the Great Pyramid of Cheops. Evidence from other pyramids built in the same period indicates that such structures were erected at incredible speeds. At Dahshur, for example, is the Pyramid of Sneferu, approximately two-thirds the volume of the Great Pyramid. An inscription in the north-east cornerstone of the structure reveals that it was laid in the 21st year of Sneferu's reign, while half-way up is a block with another inscription, dated in the 22nd year. In other words, it took only two years to raise the entire Pyramid of Sneferu.

A similar situation may also have occurred with the Cheops structure, because it was completed in as little as four years' time. The fact that recent excavations not far from the Great Pyramid have uncovered the remains of only 4000 workmen's huts increases rather than decreases the problem. There is no way 100,000 labourers could have been housed in 4000 small huts, not to mention the additional hundreds of thousands who were involved. This undoubtedly places the historians in a difficult position, for how can one explain the building of the Great Pyramid in only four years' time by just 4000 workers, if only wooden sledges and barges were utilized during a three months' period every year?

Yet it was done, and probably in just that length of time, but the builders used construction and engineering skills and techniques known only to them. It was a technological feat beyond comparison in either the ancient or the modern world. The generations following the one that built Cheops soon found themselves, however, in a steep decline. They were suffering from atrophy of knowledge, a recession in technical ability and cultural sophistication that permeated each succeeding dynasty until the Egyptian civilization became a vague shadow of its historical greatness. The hieroglyphics from the various dynasties reveal decided changes in the Egyptians' life-style and technology, and the combination of funerary texts known as the *Book of the Dead* (mentioned in Chapter 1) strongly supports this.

The Egypt we know from the history books was indeed a mere remnant of a highly progressive people who inherited technical ability beyond our understanding. The knowledge that sparked their civilization was transmitted to them by the eight survivors of the Flood, and using this knowledge, Menes, the founder of Egypt, rose to the challenge and began to transform chaos into order.

Epilogue

What really transpired on this planet in the relatively early years of human development will undoubtedly remain the subject of heated controversy for years to come. Even a detailed account of the near-unbelievable feats of prehistoric technological inventiveness still leaves it difficult for us to comprehend fully the outstanding accomplishments of our 'primitive' ancestors. Yet a thoughtful look at what the earth's crust has quietly preserved for us can enable our minds to slip back into the realm of unrecorded history and retrieve from it those minute details which increase not only our understanding but also our bewilderment, and which stimulate our desire to learn more and more and more.

There is another way to interpret history – the ooparts have proven that. The major assumption of orthodox historians – that our civilization is the result of gradual development from primitive beginnings – can now seriously be challenged. Ooparts, Biblical history, archaeology, geology, palaeontology, and ordinary level-headed thinking have guided us in that direction.

The weight of evidence is growing daily – evidence that our early ancestors created a society that surpassed ours in all aspects of development. Let's not sell humanity short by attempting to link the remains of the ancient technology to supposed visits of creatures from outer space, by ascribing to beings from other planets what in reality is the logical effect of the synergistic growth of a human super-civilization.

Our beliefs about the prehistoric ages are constantly being altered by new archaeological and palaeoanthropological findings, and thus in time significant portions of previously accepted and even of now-developing historical frameworks may become outdated and may need to be changed. The surface of historical interpretation has scarcely been scratched. Even the accumulated facts gathered in these pages should be viewed as a vehicle to stimulate deeper and more detailed studies.

An overview of history as it now appears to us may have grave implications for our future, for the world has undergone a number of important transitions, with still more to come.

Although we cannot accurately assign dates to memorable events that transpired in history, it is believed that the years between 1950 and 220 BC marked a period of transition for almost every civilization of the Old and New worlds. During this time, Egypt's first kingdom slipped into paralyzing deterioration; Sumeria and India were overwhelmed by barbaric invaders; China and the rest of the Far East suffered a disastrous flood; and in the Americas, the so-called primitive cultures were suddenly followed by more advanced ones. In many instances, the societies that collapsed and disappeared had had historical ties of one kind or another with scattered remnants of the lost super-civilization, which in turn was related to the world order of the antediluvians through the Babel world centre. The disintegration of the primary stages of the known civilizations within a relatively short time of each other at the end of the third millennium BC is historically unexplainable. No single all-embracing cause can be given for their sudden decline. The first global order was swept away by a devastating Flood; the revival of world order broke down at Babel. Both of these catastrophes destroyed order, but not the memory of the technology the ancients had once enjoyed. The terrifying means by which an oppressive authority might once again consolidate its power for world domination remained intact. Some of those who were entrusted with the preservation of this awesome knowledge eventually used it to destroy one another in a succession of nuclear holocausts. The survivors who safeguarded the secret of the great knowledge ultimately fused it with the cultures of subsequent civilizations. These civilizations lasted to the end of the third millennium BC and might have possessed sufficient potential to enable yet another global authority to threaten nuclear warfare, but too much time had elapsed and the desire for a world order had passed.

After 2000 BC, as each of the Middle Eastern civilizations experienced a brief period of revival, remnants of earlier advanced technology once again surfaced, now greatly diminished, however. Both Egypt and Babylonia seem to have preserved a number of sophisticated records and artifacts from former civilizations. The years between 250 BC and the dawn of the Christian era witnessed a technological rekindling in these lands, which produced, among other things, the electric battery used in Iraq during the Parthian period, a small computer calendar constructed in Greece in approximately 80 BC, and a model glider plane tested on the banks of the Nile during the reign of the Ptolemies.

The brutal Roman invasion of the Middle East in the first century BC extinguished this spark of revival. The Romans' ruthlessness was an integral part of a wave of wanton destruction that struck the Library of Carthage in 146 BC, reducing its irreplaceable 500,000 volumes to ashes. Later, at Pergamus in Asia Minor, another 200,000 manuscripts, known to have contained occult knowledge and perhaps the pre-Flood and pre-Babel wisdom of the occult energies, were consigned to flames by rampaging Christians. The most devastating blow, however, was dealt by Julius Caesar when he burned the athenaeum of Alexandria, destroying 700,000 of the most valued scientific works of the classical world.

The few records that survived were jealously guarded by the secret societies. Gradually these too passed into oblivion as a result of relentless persecution and mounting ignorance; as each society died, its secrets perished with it or were hidden in depositories. never to be found again.

Today we are witnessing a rebirth in science and technology, which to a large degree is a phenomenon totally independent of historical developments. The first signs of a new scientific thrust appeared in the West, primarily in Europe, and finally achieved full maturity in the Industrial Revolution. As our modern development becomes more complex and more daring, we are beginning to re-evaluate the remains and the artifacts of the past and to recognize in them plateaus of knowledge we ourselves are only now attaining.

We can wonder at this startling discovery of past accomplishments, but it must also serve as a warning. Once again, science is beginning to reach beyond the boundary separating natural science from supernatural manipulation and again we are stepping into the perilous region of the occult that was so boldly penetrated by the antediluvians and the builders. Are we once again approaching a danger point?

It has been said that history possesses the strange and unexplainable ability to repeat itself.

Will we give it the impetus to make it happen – again?

NOTES

Chapter 1

1 Johannes Riem, *Die Sintflut in Sage und Wissenschaft*, Hamburg: Agentur des Rauhen Hauses, 1925, p. 7.
2 Hugh Miller, *The Testimony of the Rocks*, New York: John B. Alden, 1892, p. 284
3 As quoted by Dr Aaron Smith, private files.
4 Lowell Thomas, *Hungry Waters – The Story of the Great Flood*, Philadelphia: John C. Winston Co., 1937, p. 184.
5 I. S. Bartlett, *History of Wyoming*, Vol. 1, Chicago: S. J. Clarke, 1918, p. 62.
6 Alexander H. Burr, *Mythology of All Races*, X, New York: Cooper Square Publishers, p. 222.
7 Alfred M. Rehwinkel, *The Flood*, St Louis: Concordia Publishing House, 1951, p. 128.
8 *ibid.*, p. 162.
9 Harold G. Coffin, *Creation – Accident or Design?*, Washington, DC: Review & Herald Publishing Association, 1969, p. 65.
10 Immanuel Velikovsky, *Earth in Upheaval*, Garden City, N.Y.: Doubleday, 1955, p. 222.
11 Miller, *The Old Red Sandstone*, Boston: Gould & Lincoln, 1857, p. 221.
12 Harry S. Ladd, 'Ecology, Paleontology and Stratigraphy', *Science*, Vol. 129, 9 January 1959, p. 32.
13 Edwin H. Colbert, *The Age of Reptiles*, New York: McGraw-Hill, 1951, p. 191.
14 Quoted in Carl Dunbar, *Historical Geology*, New York: John Wiley, 1969, p. 426.
15 Coffin, pp. 73, 74.
16 *ibid.*, p. 76.
17 S. N. Kramer, 'The Sumerians', *Scientific American*, Vol. 197, 1957, pp. 70–83.
18 H. R. Hall, 'Archaeology', *Encyclopaedia Britannica*, 1959 ed., Vol. 8, p. 37.
19 E. Wallis Budge, *The Book of the Dead: Papyrus of Ani*, New York: Dover, 1967, p. 6.
20 *ibid.*, p. 7.
21 Rene Noorbergen, private files.
22 Hans Selye, 'Is Aging Curable?', *Science Digest*, Vol. 46, December 1959, p. 1.
23 'Population Growth', *Science*, Vol. 129, 3 April 1959, p. 882.
24 William R. Vis, 'Medical Science and the Bible', *Modern Science and Christian Faith*, 2nd ed., Wheaton, Ill.: Van Kampen Press, 1950, p. 242.
25 John C. Whitcomb, Jr., and Henry M. Morris, *The Genesis Flood*, Philadelphia: Presbyterian & Reformed Publishing Co., 1961, p. 27.

26 Andrew Tomas, *We Are Not the First*, London: Souvenir Press, 1971, p. 162.
27 Campbell, *The Masks of God*, New York: Viking Press, p. 228.
28 Harry M. Orlinsky, *Ancient Israel*, Ithaca, N.Y.: Cornell University Press, 1960, pp. 6–8.
29 W. F. Albright, *Recent Discoveries in Bible Lands*, New York: Funk & Wagnalls Co., 1955, p. 4.
30 A. J. White, 'Radio Carbon Dating', *Creation Research Society Quarterly*, December 1972, pp. 156–8.
31 Robert Charroux, *Forgotten Worlds*, New York: Walker & Co., 1973, pp. 64–5.
32 Alan and Sally Landsburg, *In Search of Ancient Mysteries*, New York: Bantam Books, 1974, p. 21.
33 Louis Pauwels, *The Eternal Man*, New York: Avon, 1972, pp. 62–3.
34 Tomas, p. 8.
35 Charles Berlitz, *Mysteries from Forgotten Worlds*, New York: Dell Publishing Co., Inc., 1972, pp. 35–6.
36 Tomas, p. 8.

Chapter 3

1 Albright, p. 70.
2 John Philip Cohane, *The Key*, New York: Crown Publishers, Inc., 1970.
3 *ibid*.
4 John Michell, *The View over Atlantis*, New York: Ballantine Books, 1972, p. 69.
5 *ibid.*, p. 129.

Chapter 5

1 As quoted by Landsburg, Alan and Sally, *In Search of Ancient Mysteries*, New York: Bantam Books, 1974, p. 161.
2 R. C. Majimdar, ed., *The Vedic Age*, London: George Allen and Unwin, Ltd., 1951, p. 268.
3 Erich von Fange, 'Strange Fire on the Earth', *Creation Research Society Quarterly*, December 1975, p. 132.
4 Richard E. Mooney, *Colony Earth*, Greenwich, Connecticut: Fawcett Publications, 1974, pp. 242–3.
5 M. H. J. T. Van der Veer and P. Moerman, *Hidden Worlds*, New York: Bantam Books, 1973.

Chapter 6

1 Bjorn Kurten, *Not from the Apes*, New York: Pantheon Books, 1972, pp. 4–5.
2 William L. Straus, Jr., and A. J. A. Cave, 'Pathology and the Posture of the Neanderthal Man', *Quarterly Review of Biology*, 1957, 32: 348–63.
3 Robert Silverberg, *Man Before Adam*, Philadelphia: Macrae Smith Company, 1964, p. 191.
4 William Dick and Henry Gris, 'Delicate Head Surgery Was Per-

formed 3500 Years Ago', *National Enquirer*, 10 September 1972, p. 30.

Chapter 7

1 Gerald S. Hawkins, *Beyond Stonehenge*, New York: Harper and Row, 1973, p. 113.
2 Alexander Thoms, *Megalithic Lunar Observations*, Oxford: Clarendon Press, 1971, pp. 115–16.
3 *ibid.*

BIBLIOGRAPHY

Bailey, James, *The God-Kings and the Titans*. New York: St Martin's Press, 1973.

Baldwin, John D., *Pre-Historic Nations*. New York: Harper and Brothers, 1869.

Baring-Gould, S., *Legends of the Patriarchs and Prophets*. New York: Henry Regnery Company, 1872.

Bergier, Jacques. *Extraterrestrial Intervention: The Evidence*. Chicago: Henry Regnery Company, 1974.

—— *Extraterrestrial Visitations from Prehistoric Times to the Present*. New York: New American Library, Inc., 1974.

Berlitz, Charles. *The Bermuda Triangle*. Garden City, N.Y.: Doubleday and Company, Inc., 1974.

—— *Mysteries from Forgotten Worlds*. New York: Dell Publishing Co., Inc., 1972.

Binder, Otto. *Mankind – Colony of the Stars*. New York: Tower Publications, Inc., 1974.

—— *Unsolved Mysteries of the Past*. New York: Tower Publications, Inc., 1970.

Braidwood, Robert J. *Prehistoric Man*. New York: William Morrow & Co., 1967.

Charroux, Robert. *Forgotten Worlds*. New York: Walker and Company, 1973.

—— *The Gods Unknown*. New York: Berkeley Publishing, 1974.

—— *Legacy of the Gods*. New York: Berkeley Publishing, 1974.

—— *One Hundred Thousand Years of Man's Unknown History*. New York: Berkeley Publishing, 1971.

DeCamp, L. Sprague and Catherine C. *Ancient Ruins and Archaeology*. Garden City, N.Y.: Doubleday and Co., Inc., 1962.

Dick, William, and Henry Gris. 'Delicate Head Surgery Was Performed 3500 Years Ago'. *National Enquirer*, 10 September 1972.

Drake, W. Raymond. *Gods and Spacemen in the Ancient East*. New York: New American Library, Inc., 1973.

—— *Gods and Spacemen in the Ancient West*. New York: New American Library, Inc., 1974.

'Electric Battery of 2000 Years Ago'. *Discovery*, March 1939.

Eskridge, Robert Lee. *Mangareva: The Forgotten Islands*. Indianapolis: The Bobbs-Merrill Company, Inc., 1971.

Filby, Frederick A. *The Flood Reconsidered*. Grand Rapids, Michigan: Zondervan Publishing House, 1971.

Garvin, Richard. *The Crystal Skull*. Garden City, N.Y.: Doubleday & Co., Inc., 1973.

Gauquelin, Michel. *The Cosmic Clocks*. New York: Henry Regnery Company, 1967.

Goodavage, Joseph F. *Astrology: The Space Age Science*. New York: New American Library, Inc., 1966.

Hapgood, Charles H. *Maps of the Ancient Sea Kings*. Philadelphia: Chilton Books, 1966.
Hawkins, Gerald S. *Beyond Stonehenge*. New York: Harper and Row, 1973.
—— *Stonehenge Decoded*. New York: Dell Publishing Co., Inc., 1971.

'Job and Science'. *Five Minutes with the Bible and Science*, March–April 1973.

Keel, John. *Our Haunted Planet*. New York: Fawcett Publications, 1971.
Kingsley, Errol. *Atoms and God*. Portland, Oregon: Kingsley, 1973.
Kolosimo, Peter. *Not of This World*. New York: Bantam Books, 1973.
—— *Timeless Earth*. New York: Bantam Books, 1973.
Kurten, Bjorn. *The Ice Age*. New York: G. P. Putnam's Sons, 1972.

Landsburg, Alan and Sally. *In Search of Ancient Mysteries*. New York: Bantam Books, 1974.
Laufer, B. *The Prehistory of Aviation*. Chicago: Field Museum of Natural History, 1928.
Lissner, Ivan. *The Silent Past*. New York: G. P. Putnam's Sons, 1962.

Manning, Alan G. 'Can Pyramid Power Work for You?' *Occult*, October 1973.
Marshack, Alexander. *The Roots of Civilization*. New York: McGraw-Hill Book Company, 1972.
Maziere, Francis. *Mysteries of Easter Island*. New York: Tower Publications, Inc., 1965.
Mertz, Henriette. *Gods from the Far East*. New York: Ballantine Books, 1972.
Montgomery, John W. *The Quest for Noah's Ark*. Minneapolis: Bethany Fellowship, Inc., 1972.
Mooney, Richard E. *Colony Earth*. Greenwich, Connecticut: Fawcett Publications, 1974.

Norman, Eric. *Gods, Demons and UFO's*. New York: Lancer Books, 1970.

Ostrander, Sheila, and Lynn Schroeder. *Handbook of Psi Discoveries*. New York: G. P. Putnam's Sons, 1974.

Pauwels, Louis. *The Eternal Man*. New York: Avon Books, 1972.
—— *Impossible Possibilities*. New York: Stein and Day, Publishers, 1971.
—— *The Morning of the Magicians*. New York: Avon Books, 1968.
Peet, Eric T. *Rough Stone Monuments and Their Builders*. London: Harper and Brothers, 1912.
Price, Derek J. de Solla. 'An Ancient Greek Computer.' *Scientific American*, June 1969.

Rand, Howard B. *The World That Then Was*. Merrimac, Massachusetts: Destiny Publishers, 1954.

Roe, Derek. *Prehistory*. Berkeley, California: University ot California Press, 1970.

Salm, W. 'Babylon Battery.' *Popular Electronics*, July 1964.

Sanderson, Ivan T. *Investigating the Unexplained*. Englewood Cliffs, N.J.: Prentice-Hall, Inc., 1972.

—— *Invisible Residents*. New York: The World Publishing Company, 1970.

—— 'This Airplane Is More Than 1000 Years Old. *Argosy*, November 1969.

Silverberg, Robert. *Man Before Adam*. Philadelphia: Macrae Smith Co., 1964.

Spence, Lewis. *The History of Atlantis*. New York: University Books, 1968.

Steiger, Brad. *Atlantis Rising*. New York: Dell Publishing Co., Inc., 1973.

—— *Mysteries of Time and Space*. Englewood Cliffs. N.J.: Prentice-Hall, Inc., 1974.

Stern, Philip Van Doren. *Prehistoric Europe*. New York: W. W. Norton & Co., Inc., 1969.

Taylor, John. *The Great Pyramid*. London: Longman, Green, Longman and Roberts, 1859.

Thoms, A. *Megalithic Lunar Observations*. Oxford: Clarendon Press, 1971.

—— *Megalithic Sites in Britain*. Oxford: Clarendon Press, 1971.

Tomas, Andrew. *We Are Not the First*. New York: Bantam Books, 1973.

Tompkins, Peter. *Secrets of the Great Pyramid*. New York: Harper and Row, 1971.

Van der Veer, M. H. J. T., and P. Moerman. *Hidden Worlds*. New York: Bantam Books, 1973.

Velikovsky, Immanuel. *Earth in Upheaval*. New York: Dell Publications, Inc., 1955.

Von Daniken, Erich. *Chariots of the Gods?* New York: Bantam Books, 1971.

—— *Gods from Outer Space*. New York: G. P. Putnam's Sons, 1970.

—— *The Gold of the Gods*. New York: Bantam Books, 1974.

—— *In Search of Ancient Gods*. New York: G. P. Putnam's Sons, 1974.

Von Fange, Erich. 'Strange Fire on the Earth.' *Creation Research Society Quarterly*, December 1975.

Watson, Lyell, *Super Nature*. Garden City, N.Y.: Anchor Press, 1973.

Wernick, Robert. 'Danubian Minicivilization Bloomed Before Ancient Egypt and China.' *Smithsonian*, March 1975.

Wilson, Clifford. *Crash Go the Chariots*. New York: Lancer Books, 1972.

200

Book Tokens

**Give them
the pleasure of choosing**

Book Tokens can be bought
and exchanged at most
bookshops.

NEL BESTSELLERS

NEL P.O. BOX 11, FALMOUTH TR10 9EN, CORNWALL

Postage charge:

U.K. Customers. Please allow 25p for the first book plus 10p per copy for each additional book ordered to a maximum charge of £1.05 to cover the cost of postage and packing, in addition to cover price.

B.F.P.O. & Eire. Please allow 25p for the first book plus 10p per copy for the next 8 books, thereafter 5p per book, in addition to cover price.

Overseas Customers. Please allow 40p for the first book plus 12p per copy for each additional book, in addition to cover price.

Please send cheque or postal order (no currency).

Name ....

Address ....

....

Title ....